Numbers

Put Big Numbers into Columns and Parts

First, you need to know the names of all the columns.
For example, for the number 1 321 648:

1	3	2	1	6	4	8
Millions	Hundred Thousands	Ten Thousands	Thousands	Hundreds	Tens	Units

You can then split any number up into parts like this:

1 000 000	One million
300 000	Three hundred thousand
20 000	Twenty thousand
1 000	One thousand
600	Six hundred
40	Four tens (forty)
8	Eight units

These add together to make 1 321 648.

Look at Big Numbers in Groups of Three

To read or write a big number, follow these steps:

1) Start from the right-hand side of the number.

2) Moving left, put a space or a comma every three digits to break it up into groups of three.

3) Now going right, read each group as a separate number, as shown.

1 321 648
Millions Thousands The rest

So read as: 1 million, 321 thousand, 648.
Or write fully in words: One million,
three hundred and twenty-one thousand,
six hundred and forty-eight.

Functional Skills
Maths
Edexcel – Level 1

This brilliant CGP book is the best way to prepare for Edexcel Level 1 Functional Skills Maths. It covers everything you need... and nothing you don't!

Every topic is clearly explained, along with all the non-calculator methods you'll need for the latest exam. There are practice questions throughout the book <u>and</u> a realistic practice paper at the end — all with answers included.

How to access your free Online Edition

This book includes a free Online Edition to read on your PC, Mac or tablet.
You'll just need to go to **cgpbooks.co.uk/extras** and enter this code:

3925 1610 9317 2254

By the way, this code only works for one person. If somebody else has used this book before you, they might have already claimed the Online Edition.

CGP — still the best! ☺

D1439395

Our sole aim here at CGP is to produce the highest quality books — carefully written, immaculately presented and dangerously close to being funny.

Then we work our socks off to get them out to you
— at the cheapest possible prices.

Contents

Published by CGP

Editors:
Adam Bartlett, Michael Bushell, Tom Miles, Caley Simpson, Michael Weynberg

With thanks to Kevin Bennett and Liam Dyer for the proofreading.

ISBN: 978 1 78908 391 0

Printed by Elanders Ltd, Newcastle upon Tyne.
Clipart from Corel®

Finding the Lowest Number

1) First, find the numbers with the fewest digits.

2) From these, the one with the lowest first digit is the lowest number.

3) If the first digits are the same, look at the other digits in turn. For example, next look at the second digit and only keep the numbers with the lowest digit.

EXAMPLE:

4 people compare how many miles their cars have done.
Jeremy's car has done 81 638 miles, Grier's has done 44 653 miles,
Maggie's has done 106 464 miles, and Fraser's has done 46 150 miles.
Whose car has done the least number of miles?   *Least, smallest and fewest just mean lowest.*

1) Find the numbers with the fewest digits:

   81 638, 44 653 and 46 150 ← *These all have five digits.*

2) Find the numbers with the lowest first digit:

   44 653 and 46 150 ← *4 (in 44 653 and 46 150) is lower than 8 (in 81 638).*

3) Find the number with the lowest second digit:

   44 653 ← *4 (in 44 653) is lower than 6 (in 46 150).*

So **Grier's** car has done the least number of miles.

Finding the Highest Number

1) First, find the numbers with the most digits.

2) From these, the one with the highest first digit is the highest number.

3) If the first digits are the same, look at the other digits in turn. For example, next look at the second digit and only keep the numbers with the highest digit.

EXAMPLE:

Find the highest number from this list:

158 573,   3147,   215 608,   5410,   4683,   52 010

1) Find the numbers with the most digits:

   158 573,   215 608 ← *These both have six digits.*

2) Find the number with the highest first digit:

   215 608 ← *The first digit of 215 608 is 2, which is higher than 1 (in 158 573).*

So **215 608** is the highest number.

Putting Whole Numbers in Order

1) First, sort the numbers into groups with the same number of digits.
   Numbers with more digits are larger and numbers with fewer digits are smaller.

2) Then put each group in order. Start by looking at the first digit.

3) If the first digits are the same, move onto the second digit, then the third digit, and so on.

EXAMPLE:

Put these numbers in order from largest to smallest.
29 635, 140 246, 301 564, 468, 894, 93 645, 4545, 168

1) Put the numbers into groups. The ones with the most digits should go into the first group.

6 digits	5 digits	4 digits	3 digits
140 246, 301 564	29 635, 93 645	4545	468, 894, 168

2) Put each separate group in order of size from largest to smallest.

6 digits	5 digits	4 digits	3 digits
301 564, 140 246	93 645, 29 635	4545	894, 468, 168

3) So the numbers in order from largest to smallest are:

301 564, 140 246, 93 645, 29 635, 4545, 894, 468, 168

Practice Questions

1) Write the number 503 425 in words.

....

2) Look at these numbers: 3457, 165, 671 902, 54 927, 210, 16 324, 275 769, 216.

   a) Find the lowest number.   b) Find the highest number.

   ....

3) The list below shows how much money six people spent on their holidays last year.

Neil	£2415	Yuri	£13 420
Amelia	£15 432	Marco	£546
Rosa	£5246	Marie	£552

Put the names in order from the person who spent the most money on their holidays to the person who spent the least money on their holidays.

....

The Number Line and Scales

Negative Numbers are Less than Zero

1) A negative number is a number less than zero.

2) You write a negative number using a minus sign (–). For example, –1, –2, –3.

3) A number line is really useful for understanding negative numbers.

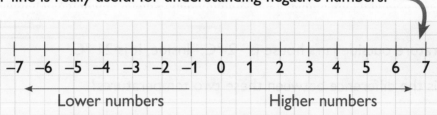

-7 -6 -5 -4 -3 -2 -1 0 1 2 3 4 5 6 7

← Lower numbers   Higher numbers →

All negative numbers are to the left of zero.

All positive numbers are to the right of zero.

The further right you go, the higher the numbers get.
For example, –2 is higher than –7.

Use a Number Line to Work Out Differences

You can use a number line to work out the difference between two numbers.
For example, the difference between a negative number and a positive number.

EXAMPLE:

In London, the temperature is –2 °C. In Paris, it's 6 °C.

What is the difference in temperature between the two places?

1) Draw a number line that includes both the numbers in the question.

2) Count on from –2 to 6.

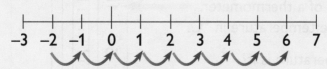

-3 -2 -1 0 1 2 3 4 5 6 7

There are 8 steps, so the difference in temperature is **8 °C**.

Practice Questions

1) Which of these numbers is lower: –8 or –4?

....

2) In the morning, it is –3 °C. At night, it is –5 °C.
   Is the temperature higher in the morning or at night?

....

3) The temperature outside Jan's house is –2 °C. Inside, it is 21 °C.

   a) What is the difference between these two temperatures?

   ....

   ....

   b) Overnight, the outside temperature falls to –6°C. The inside temperature
      stays the same. What is the difference between the temperatures now?

   ....

   ....

A Scale is a Type of Number Line

1) You might be asked to read a scale.
   For example, to read the temperature off a thermometer.

2) Scales are just number lines.
   They don't always show every number though.

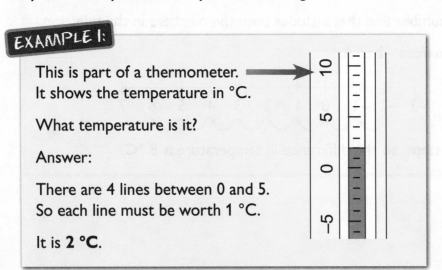

EXAMPLE 1:

This is part of a thermometer.
It shows the temperature in °C.

What temperature is it?

Answer:

There are 4 lines between 0 and 5.
So each line must be worth 1 °C.

It is **2 °C**.

EXAMPLE 2:

This is part of a thermometer.
It shows the temperature in °C.

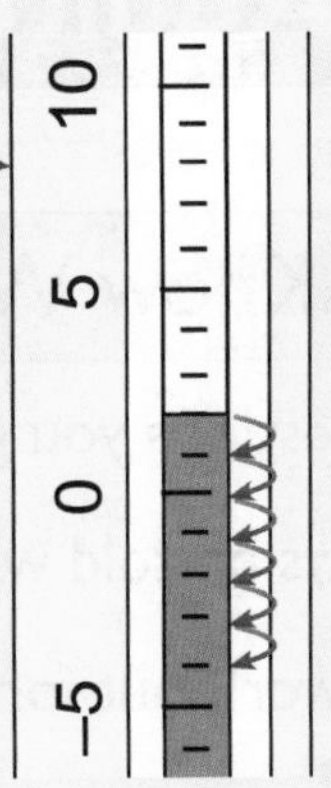

What would the thermometer read if
the temperature dropped by 6 °C?

Answer:

1) Work out what the thermometer reads now.

2) Count down 6 places.

The thermometer would read **–4 °C**.

EXAMPLE 3:

This is part of a thermometer.
It shows the temperature in °C.

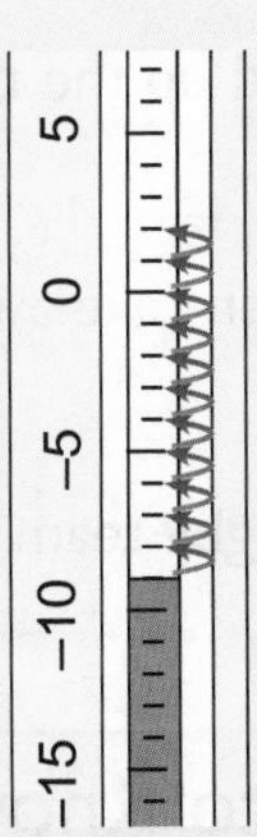

What would the thermometer read if
the temperature increased by 11 °C?

Answer:

1) Work out what the thermometer reads now.

2) Count up 11 places.

The thermometer would read **2 °C**.

Practice Questions

Part of a thermometer is shown on the right. It gives the temperature in °C.

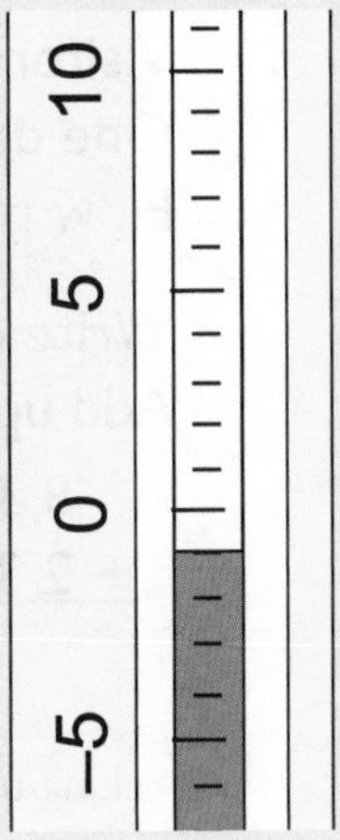

1) What temperature is shown on the thermometer?

....

2) What temperature would this thermometer read
   if the temperature dropped by 4 °C?

....

3) Looking again at the thermometer reading on the right, what
   temperature would it read if the temperature increased by 13 °C?

....

Addition and Subtraction

You Need to Know When to Add or Subtract

1) Most of the questions you get in the test will be based on real-life situations.

2) You won't always be told whether to add or subtract (take away).

3) You'll need to work out for yourself what calculation to do.

EXAMPLE:

A rugby team wins a match. They scored 25 points from tries, 8 points from conversions and 12 points from kicks. Their opponents scored 32 points in total. How many points did the team win by?

Answer: This calculation has two steps.

1) Add up the total number of points the team scored:

$$\boxed{2}\boxed{5}\boxed{+}\boxed{8}\boxed{+}\boxed{1}\boxed{2}\boxed{=}\quad \boxed{45}$$

2) Then take away the points their opponents scored:

$$\boxed{4}\boxed{5}\boxed{-}\boxed{3}\boxed{2}\boxed{=}\quad \boxed{13}$$

The rugby team won by **13** points.

You Need to Know How to Add Without Using a Calculator

1) Write the numbers with one above the other, lining up the units columns.

2) Add up the columns from right to left. Write the sum at the bottom of each column.

3) If the digits add up to more than 9, write the units digit at the bottom of the column and 'carry' the 1 to the next column (write a little 1 at the bottom of the next column).

EXAMPLE:

Visitors arrive at a theme park by car or by coach.
One day, 8340 people arrived by coach and 2378 arrived in cars.
How many people visited the theme park in total that day?

Write out the numbers one above the other, lining up the units columns.
Add up each column in turn, starting with the units.

```
  8 3 4 0        8 3 4 0        8 3 4 0        8 3 4 0        8 3 4 0
+ 2 3 7 8      + 2 3 7 8      + 2 3 7 8      + 2 3 7 8      + 2 3 7 8
                        8           1 8          7 1 8      1 0 7 1 8
                                      1            1            1
```

Line up the units columns.

$0 + 8 = 8$

$4 + 7 = 11$ Write 1 and 'carry' 1.

$3 + 3 + $ carried $1 = 7$

$8 + 2 = 10$ There are no more columns left so you don't need to carry. You can just write 10.

So the total number of visitors is **10 718**.

You Need to Know How to Subtract Without Using a Calculator

1) Write the numbers with the bigger one on top, lining up the units columns.

2) Start with the units column. Take the bottom digit away from the top digit.

3) If the top digit is smaller than the bottom digit, you need to borrow ten from the next column along. Subtract 1 from the column to the left and add 10 to the column you're subtracting.

4) Now do the same for each column in turn from right to left.

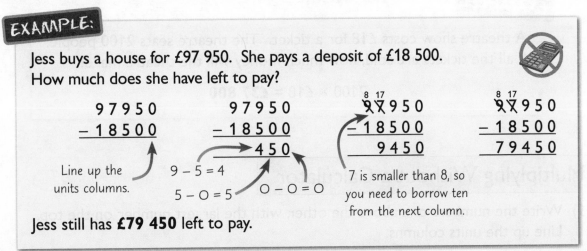

EXAMPLE:

Jess buys a house for £97 950. She pays a deposit of £18 500.
How much does she have left to pay?

$$
\begin{array}{r}
9\,7\,9\,5\,0 \\
-\,1\,8\,5\,0\,0 \\
\end{array}
$$

Line up the units columns.

$$
\begin{array}{r}
9\,7\,9\,5\,0 \\
-\,1\,8\,5\,0\,0 \\
\hline
4\,5\,0 \\
\end{array}
$$

9 − 5 = 4
5 − 0 = 5
0 − 0 = 0

$$
\begin{array}{r}
{}^{8}\cancel{9}\,{}^{17}\cancel{7}\,9\,5\,0 \\
-\,1\,8\,5\,0\,0 \\
\hline
9\,4\,5\,0 \\
\end{array}
$$

7 is smaller than 8, so you need to borrow ten from the next column.

$$
\begin{array}{r}
{}^{8}\cancel{9}\,{}^{17}\cancel{7}\,9\,5\,0 \\
-\,1\,8\,5\,0\,0 \\
\hline
7\,9\,4\,5\,0 \\
\end{array}
$$

Jess still has **£79 450** left to pay.

Practice Questions

1) On Monday, Alice did 17 650 steps. On Tuesday, she did 9820 steps.
How many steps did Alice do in total on Monday and Tuesday?

....

2) Amir is baking a wedding cake. He has 3200 g of sugar in his cupboard.
The recipe says that he will use 2450 g of sugar. How much sugar will Amir have left over?

....

3) A company has £50 430 in its bank account. They buy a new van for £34 650.
How much money will the company have left?

....

Multiplication and Division

You Need to Know When to Multiply

1) Some calculations will involve multiplication — one number "times" another.

2) Multiplication is shown using a × sign.

> **EXAMPLE:**
>
> A theatre show costs £18 for a ticket. The theatre seats 2100 people.
> If all the tickets are sold, how much money will the theatre make?
>
> 2100 × £18 = **£37 800**

Multiplying Without a Calculator

1) Write the numbers one above the other, with the largest number on the top.
   Line up the units columns.

2) Multiply the units digit of the bottom number by each digit of the top number in turn
   (working from right to left). Write the answer underneath.

3) Then multiply the tens digit of the bottom number by each digit of the top number.
   You need to add a zero in the units column of the answer as you're multiplying by ten.

4) Remember, if you get an answer of **10** or more, carry the tens digit of the answer to the
   next column to the left (like you do when you're adding). Add any carried numbers after
   doing the multiplication for the next column.

5) Finally, add your answers together to get the final answer.

> **EXAMPLE:**
>
> Adam buys a new TV. He has to pay £106 a month for 18 months.
> How much will Adam pay in total for the TV?
>
>
>
> **1) First find 106 × 8.**
>
> ```
>       1 0 6
>     ×   1 8
>  8×1=8 8 4 8
>         4
> ```
>
> 8 × 6 = 48,
> so put 8 in the
> units column and
> carry the 4 to
> the tens column.
>
> 8 × 0 = 0,
> 0 + carried 4 = 4,
> so put 4 in the tens column.
>
> **2) Then find 106 × 1.**
>
> ```
>     1 0 6
>   ×   1 8
>     8 4 8
>   1 0 6 0
> ```
>
> When you're multiplying by the tens
> digit you need to write a 0 in the
> units column. All the other digits
> are shifted one column to the left.
>
> **3) Add to get the answer.**
>
> ```
>     1 0 6
>   ×   1 8
>     8 4 8
> + 1 0 6 0
>   1 9 0 8
>       1
> ```
>
> See page 8
> for more on
> non-calculator
> addition.
>
> Adam will pay **£1908** in total for the TV.

You Need to Know When to Divide

1) Some calculations will involve division — one number divided by another.

2) Division is shown using a ÷ sign.

EXAMPLE:

A group is budgeting for a 5-day holiday.
How much money does the group have per day if they have £1360 for the 5 days?

Answer: £1360 needs to be divided by 5 days.
So you need to calculate 1360 divided by 5.

$$1360 \div 5 = 272$$

So the group has **£272** per day to spend on the holiday.

Using the non-calculator method (see below), you would do:

```
   0 2 7 2
5 |1 ¹3 ³6 ¹0
```

Dividing Without a Calculator

1) Write the small number to the left of the big number.

672 ÷ 4 would be written like this:

$$4 \,|\, 6 \; 7 \; 2$$

2) Divide each digit of the bigger number by the small number, working from left to right. Put the result of each division above the big number, lining up the digits.

3) If the number doesn't divide exactly, then the number left over after the division is carried over to the next column. It's then used in the next digit's division (so if a 1 was carried over and the next digit was a 6, 16 would be used for the next division).

4) The amount left over after all the digits have been divided is called the remainder.

EXAMPLE:

Kofi has raised £1570 for charity. He wants to split it equally between 9 charities. How much money will each charity get and how much will be left over?

Answer: You need to divide 1570 by 9.
Write the division with the small number on the left: 9 |1 5 7 0

```
    0
9 |1 ¹5 7 0
```
9 doesn't divide 1 exactly, so write a 0 on the top line and carry over the 1.

```
    0 1
9 |1 ¹5 ⁶7 0
```
9 goes into 15 once with 6 left over. Write a 1 on the top line and carry over the 6.

```
    0 1 7
9 |1 ¹5 ⁶7 ⁴0
```
9 goes into 67 seven times with 4 left over. Write a 7 on the top line and carry over the 4.

```
    0 1 7 4 r 4
9 |1 ¹5 ⁶7 ⁴0
```
9 goes into 40 four times with 4 left over. Write a 4 on the top line. 4 is left over as a remainder.

So each charity will get **£174** and there will be **£4** left over.

Some Questions Need Answers that are Whole Numbers

1) You won't always end up with a whole number when you divide.

2) But sometimes, you'll need to give a whole number as your answer.

> **EXAMPLE:**
>
> Richard needs 435 burgers for a summer event.
> Burgers come in packs of 12. How many packs will he need?
>
> Calculation: $435 \div 12 = 36.25$ ← Without a calculator, you'd get $435 \div 12 = 36$ r 3.
>
> Richard can't buy 36.25 packs, so your answer needs to be a whole number.
>
> 36.25 is between 36 and 37. If Richard buys 36 packs of burgers, he won't have enough. So Richard will need to buy **37 packs**.

Practice Questions

1) Arwa is organising a trip. She has booked 5 coaches. Each coach seats 48 people. How many people can go on Arwa's trip?

....

2) Jerome sells his jam tarts in packs of 8. He has **184** jam tarts. How many packs of jam tarts can he make altogether?

....

3) Stephanie sells chocolates in boxes of 9.

a) One day, she makes 175 chocolates. How many boxes of chocolates can she make and how many chocolates will she have left over?

....

b) The next day she makes enough chocolates to fill **123** boxes with 7 chocolates left over. How many chocolates does she make on this day?

....

Checking Your Answer

Always Check Your Answer

1) Once you've got your answer, you should check it using the opposite calculation.

2) Adding and subtracting are opposite calculations.

3) Multiplication and division are also opposite calculations.

4) By doing the opposite calculation you should get back to the number you started with.

EXAMPLE 1:

What is $300 - 102$?

Answer: $300 - 102 = \textbf{198}$

Check it using the opposite calculation, so you add: $198 + 102 = 300$

EXAMPLE 2:

What is 36×54?

Answer: $36 \times 54 = \textbf{1944}$

Check it using the opposite calculation, so you divide:

$1944 \div 54 = 36$ OR $1944 \div 36 = 54$

You only need to do one of these calculations to check your answer.

Practice Questions

1) $11 \times 33 = 363$
   Show how you could check whether or not this calculation is correct.

   ....

2) Mark climbs up two mountains. One is 803 m high and the other is 899 m high.
   He calculates that he has climbed 1602 m. Check if Mark's calculation is correct.

   ....

3) Allison is making party bags. She has 144 sweets and wants to put 6 in each bag.
   She calculates that she can make 24 party bags. Check if Allison's calculation is correct.

   ....

Multiplying and Dividing by 10, 100, 1000, etc.

Multiplying Any Number by 10 or 100

1) To multiply by 10, move the decimal point one place to the right.

2) To multiply by 100, move the decimal point two places to the right.

3) If you need to, add zeros on the end to fill any gaps.

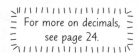
For more on decimals, see page 24.

33 × 10 = 3 3 0

9.1 × 10 = 9 1

It doesn't look like there's a decimal point in 33 or 65, but remember that 33 = 33.0 and 65 = 65.0.

65 × 100 = 6 5 0 0

2.3 × 100 = 2 3 0

Multiplying Any Number by 1000, etc.

1) Count the number of zeros in the number you're multiplying by. Then move the decimal point that number of places to the right.

2) If you need to, add zeros on the end.

49 × 1000 = 4 9 0 0 0

2.59 × 1000 = 2 5 9 0

1.389 × 10 000 = 1 3 8 9 0

You always move the decimal place this much:

| 1 place for 10 | 3 places for 1000 |
| 2 places for 100 | 4 for 10 000 etc. |

Practice Questions

1) Work out:

   a) 23 × 10

   b) 4.65 × 10

   c) 7.89 × 100

   d) 97 × 100

   ....

2) Work out:

   a) 65.7 × 1000

   b) 0.46 × 1000

   c) 0.978 × 10 000

   d) 29.8 × 100 000

   ....

3) Mount Everest is 10 times as high as Esk Pike. Esk Pike is 885 m high.
   How high is Mount Everest?

   ....

Dividing Any Number by 10 or 100

1) To divide by 10, move the decimal point one place to the left.

2) To divide by 100, move the decimal point two places to the left.

3) If you need to, remove zeros from the end.

$$330 \div 10 = 3\,3\,\cancel{0} = 3\,3$$
$$91 \div 10 = 9.1$$

$$6500 \div 100 = 6\,5\,\cancel{0}\,\cancel{0} = 6\,5$$
$$213.7 \div 100 = 2.1\,3\,7$$

Dividing Any Number by 1000, etc.

1) Count the number of zeros in the number you're dividing by.
   Then move the decimal point that number of places to the left.

2) If you need to, remove zeros from the end.

$$49\,000 \div 1000 = 4\,9\,\cancel{0}\,\cancel{0}\,\cancel{0} = 4\,9$$
$$259.6 \div 1000 = 0.2\,5\,9\,6$$
$$138\,970 \div 10\,000 = 1\,3.8\,9\,7\,\cancel{0}$$
$$= 1\,3.8\,9\,7$$

You always move the decimal place this much:

1 place for 10	3 places for 1000
2 places for 100	4 for 10 000 etc.

13.8970 is the same as 13.897 — you can remove the zero since it's the last digit.

Practice Questions

1) Work out:

   a) 18 ÷ 10

   b) 4.8 ÷ 10

   c) 46 580 ÷ 100

   d) 972 ÷ 100

   ....

2) Work out:

   a) 4159 ÷ 1000

   b) 24 ÷ 1000

   c) 32 140 ÷ 10 000

   d) 267 800 ÷ 100 000

   ....

3) A 10-person team competes in a 24-hour cycling race. In total, they cover 443 miles.
   Each person cycles the same distance. How far did each person cycle?

   ....

Square Numbers

You Need to Know How to Square a Number

1) To square a number, you just multiply the number by itself.

> 7 squared = 7 × 7 = 49   15 squared = 15 × 15 = 225

2) You can write squared numbers using a little 2 like this: ^{2}

> 7 squared can be written as 7^2 15 squared can be written as 15^2

3) If your calculator has a button like $\boxed{x^2}$, you can use it to square a number.

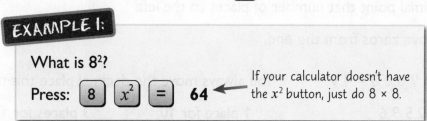

EXAMPLE 1:

What is 8^2?

Press: $\boxed{8}$ $\boxed{x^2}$ $\boxed{=}$ **64** ← If your calculator doesn't have the x^2 button, just do 8 × 8.

4) You can find a square number without a calculator. Just use the method for multiplication and multiply the number by itself.

See page 10 for how to multiply without a calculator.

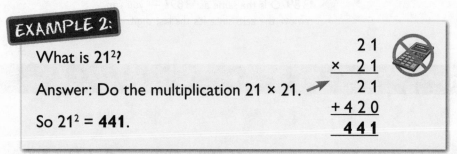

EXAMPLE 2:

What is 21^2?

Answer: Do the multiplication 21 × 21.

So 21^2 = **441**.

```
      2 1
  ×   2 1
      2 1
  + 4 2 0
    4 4 1
```

Practice Questions

1) What is:

   a) six squared?   b) 3^2?   c) 12^2?

   ....

2) Calculate 23^2.

Order of Operations

Calculations with Several Steps

1) You'll sometimes need to do calculations that have several steps.

2) You need to be careful about the order that you do a calculation.
   If you do it in the wrong order, you might get the wrong answer.

Order of Calculations (BIDMAS)

1) BIDMAS tells you the order in which calculations should be done:

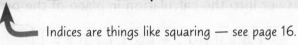

Brackets, Indices, Division, Multiplication, Addition, Subtraction

Indices are things like squaring — see page 16.

2) Work out brackets first, then indices, then multiply and divide,
   then add and subtract.

3) For multiplying and dividing, you work from left to right.
   You also work from left to right when adding and subtracting.

EXAMPLE 1:

Work out $4 + 6 \div 2$.

1) Follow BIDMAS — do the division first.

$$4 + 6 \div 2 = 4 + 3$$

2) Then do the addition.

$$4 + 3 = \mathbf{7}$$

If you didn't follow BIDMAS, you'd do
$4 + 6 \div 2 = 10 \div 2 = 5$ (which is wrong).

EXAMPLE 2:

Find $(8 - 2) \times (3 + 4)$.

1) Follow BIDMAS — start by working out what's in the brackets.

$$(8 - 2) \times (3 + 4) = 6 \times 7$$

2) Work out the multiplication.

$$6 \times 7 = \mathbf{42}$$

If you didn't follow BIDMAS, you might wrongly do
$(8 - 2) \times (3 + 4) = 6 \times (3 + 4) = 18 + 4 = 22$.

Order of Calculations Using a Calculator

1) Calculators follow BIDMAS.

2) Be careful when typing a calculation into your calculator. If you get it in the wrong order, your calculator will give the wrong answer.

3) If your calculator has bracket buttons, make sure you use them. The calculator will work out the bits inside the brackets before it does the rest of the calculation.

4) If your calculator doesn't have bracket buttons, just work out the bits in brackets for yourself first, then enter the result into your calculator.

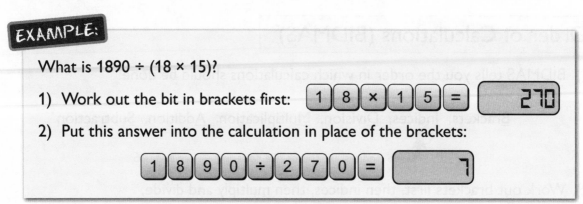

EXAMPLE:

What is $1890 \div (18 \times 15)$?

1) Work out the bit in brackets first: $\boxed{1}\boxed{8}\boxed{\times}\boxed{1}\boxed{5}\boxed{=}\ \boxed{270}$

2) Put this answer into the calculation in place of the brackets:

$\boxed{1}\boxed{8}\boxed{9}\boxed{0}\boxed{\div}\boxed{2}\boxed{7}\boxed{0}\boxed{=}\ \boxed{7}$

Practice Questions

1) What is $4 + 2 \times 5$?

....

2) Calculate:

a) $2 \times 6 - 3 \times 2$ b) $2 \times (6 - 3) \times 2$

....

3) What is $(36 \div 6) \div (24 \div 12)$?

....

4) What is $5^2 - 4 \times 6$?

....

5) Work out:

a) $65 \div 5 + 8$ b) $65 \div (5 + 8)$

....

Fractions

Fractions Show Parts of Things

1) If something is divided up into equal parts, you can show it as a fraction.

2) There are two bits to every fraction:

The bottom number shows how many parts there are in total.

$$\dfrac{3}{7}$$

The top number shows how many parts you're talking about.

EXAMPLE:

Gemma has 7 squares of chocolate. She eats 4 squares. What fraction did she eat?

She's eaten 4 out of the 7 squares, so it's $\dfrac{4}{7}$ (you say 'four sevenths').

Learn How to Write Fractions

Here's how to write some common fractions:

One half = $\dfrac{1}{2}$ One quarter = $\dfrac{1}{4}$ Three quarters = $\dfrac{3}{4}$

One third = $\dfrac{1}{3}$ One fifth = $\dfrac{1}{5}$ One tenth = $\dfrac{1}{10}$

Practice Questions

1) Jean's cat has 5 kittens. Jean gives 3 of the kittens away. What fraction of the kittens did Jean give away?

....

2) Klaus is putting up 8 shelves. He has already put up 5 shelves. What fraction of the shelves are still left to put up?

....

Mixed Numbers

1) Mixed numbers are when you have a whole number part and a fraction part together.

$1\frac{1}{4}$ $2\frac{3}{5}$

one and one quarter   two and three fifths

2) To convert a mixed number into a fraction, you first need to find the new top number.
Multiply the whole number part by the bottom number of the fraction part.
Add this to the top number of the fraction to get the new top number.

3) The bottom number stays the same.

EXAMPLE 1:

What fraction is equal to $3\frac{2}{7}$?

1) Find the new top number. Multiply the whole number part by the bottom number of the fraction. Then add this to the top number.

$3 \times 7 = 21$, so the new top number $= 21 + 2 = 23$

2) The bottom number stays the same, so $3\frac{2}{7} = \frac{23}{7}$.

4) To convert from a fraction to a mixed number, you need to divide the top number by the bottom number. You should do this without a calculator so you get a remainder.

5) The whole number part of the mixed number will be the result of the division.
The fraction part will have the remainder on the top. The bottom number stays the same.

EXAMPLE 2:

What is $\frac{13}{5}$ written as a mixed number?

1) Do the division: $13 \div 5 = 2$ remainder 3

2) The whole number is 2 and the top part of the fraction is 3, so the mixed number is $2\frac{3}{5}$.

Practice Questions

1) Convert the following mixed numbers to fractions.

a) $1\frac{4}{5}$ b) $2\frac{1}{6}$ c) $3\frac{5}{7}$ d) $2\frac{7}{10}$

....

2) Convert the following fractions to mixed numbers.

a) $\frac{5}{3}$ b) $\frac{15}{4}$ c) $\frac{18}{5}$ d) $\frac{21}{8}$

....

Equivalent Fractions

1) Equivalent fractions are equal in size, but the numbers on the top and bottom are different.

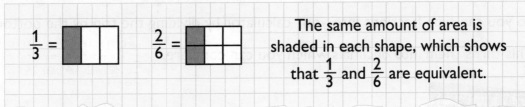

$\frac{1}{3} =$ The same amount of area is shaded in each shape, which shows that $\frac{1}{3}$ and $\frac{2}{6}$ are equivalent.

$\frac{2}{6} =$

2) To get from one fraction to an equivalent one, multiply or divide the top AND the bottom of the fraction by the same number.

EXAMPLE:

What fraction is equivalent to $\frac{3}{4}$ but has 16 on the bottom?

Answer: To get from 4 to 16, you multiply by 4. So to find the new top number, you need to multiply the top by 4 as well.

$\times 4$

New top number = 3 × 4 = 12, so $\frac{3}{4} = \frac{12}{16}$.

$\times 4$

Comparing Fractions

1) Fractions are just numbers, so they can be put in order of size.

2) Look at the bottom numbers. If they're the same, then you can just order the fractions by looking at the top numbers.

3) If the bottom numbers are different, you'll need to use equivalent fractions. Turn the fractions into equivalent fractions with the same bottom number.

4) Then you can use the top numbers to order them.

EXAMPLE:

Which is bigger, $\frac{1}{3}$ or $\frac{2}{5}$?

1) The bottom numbers are different so you need to find equivalent fractions.

2) 3 × 5 = 15, so multiply the top and bottom of each fraction to get 15 on the bottom.

$\times 5$ $\qquad$ $\times 3$

$\frac{1}{3} = \frac{5}{15}$ $\qquad$ $\frac{2}{5} = \frac{6}{15}$

$\times 5$ $\qquad$ $\times 3$

3) Now compare the top numbers — 6 is bigger than 5, so $\frac{2}{5}$ is bigger than $\frac{1}{3}$.

5) To compare mixed numbers, look at the whole number parts first. If the whole number parts are the same, then compare the fraction parts using the method above.

Comparing Fractions With a Calculator

1) $\frac{3}{4}$ is just another way of writing $3 \div 4$.

2) So you can type fractions into your calculator by dividing the top by the bottom.

3) This turns them into decimals which you can compare.

> For more on comparing decimals, see page 24.

EXAMPLE:

Put the following fractions in order from smallest to largest: $\frac{3}{5}, \frac{1}{2}, \frac{3}{4}$

$\frac{3}{5} = 3 \div 5 = 0.6$ $\qquad$ $\frac{1}{2} = 1 \div 2 = 0.5$ $\qquad$ $\frac{3}{4} = 3 \div 4 = 0.75$

So the fractions in order are $\frac{1}{2}, \frac{3}{5}, \frac{3}{4}$. $\qquad$ Make sure to write your answer as fractions, not decimals.

Practice Questions

1) What number should go in the box so that $\frac{\square}{24}$ is equivalent to $\frac{5}{8}$?

....

2) What number should go in the box so that $\frac{\square}{3}$ is equivalent to $\frac{16}{12}$?

....

3) Which fraction is larger, $\frac{3}{8}$ or $\frac{10}{24}$?

....

4) Put the following fractions and mixed numbers in order from smallest to largest.

a) $\frac{4}{8}, \frac{5}{8}, \frac{6}{10}$ $\qquad\qquad\qquad\qquad$ b) $1\frac{1}{5}, 1\frac{1}{4}, \frac{19}{20}$

.... $\qquad\qquad$ ....

.... $\qquad\qquad$ ....

5) Samina and Rachel are running a race. Samina has run $\frac{4}{9}$ of the race and Rachel has run $\frac{17}{36}$ of the race. Who has run further in the race?

....

....

'Of' means 'Times'

1) Sometimes, you might need to calculate a 'fraction of' something.
   In these cases, 'of' means 'times' (multiply).

2) To multiply a number by a fraction, you need to divide by the
   bottom number of the fraction and multiply by the top number.

EXAMPLE 1:

What is $\frac{1}{4}$ of 40?

1) 'Of' means 'times' (×), so $\frac{1}{4}$ of 40 is the same as $\frac{1}{4}$ × 40.

2) To multiply by a fraction, divide by the bottom and multiply by the top.
   So the calculation you need to do is: 40 ÷ 4 × 1 = **10**

EXAMPLE 2:

A committee with 32 members takes a vote. Three quarters of the
committee members vote 'yes'. How many members vote yes?

You need to calculate three quarters of 32.

1) 'Of' means 'times' (×), so $\frac{3}{4}$ of 32 is the same as $\frac{3}{4}$ × 32.

2) To do the calculation without a calculator, you need to divide by the
   bottom (4) and times by the top (3):   32 ÷ 4 = 8, then 8 × 3 = **24**

Practice Questions

1) What is $\frac{1}{3}$ of 18?

....

2) What is $\frac{2}{5}$ of 50?

....

3) Kieran is a hotel receptionist. He works forty-two hours a week. Kieran estimates that he
   spends a quarter of his time on the phone and a third of his time dealing with complaints.

   a)   How many hours a week does Kieran spend dealing with complaints?

   ....

   b)   How many hours a week does Kieran spend on the phone?

   ....

Decimals

Not All Numbers Are Whole Numbers

1) Decimals are numbers with a decimal point (.) in them. For example, 0.5, 1.3.

2) They're used to show the numbers in between whole numbers.

> The number 3.95 is a bit smaller than the number 4.
>
> The number 3.5 is exactly halfway between the numbers 3 and 4.
>
> The number 3.257 is slightly bigger than 3.25.

3) The first digit after a decimal point shows tenths, the second digit shows hundredths, and the third digit shows thousandths.

4) You can show decimals on a number line:

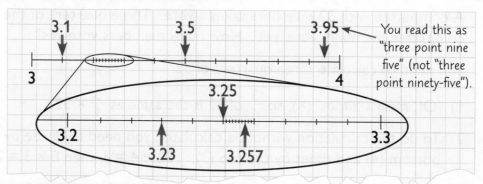

You read this as "three point nine five" (not "three point ninety-five").

How to Put Decimals in Order

You might need to arrange a list of decimal numbers in order of size.

EXAMPLE:

Put these decimals in order of size from smallest to largest: 1.7, 0.75, 0.758, 0.078

1) Put the numbers into a column, lining up the decimal points.

2) Make all the numbers the same length by filling in extra zeros at the ends.

3) Look at the numbers before the decimal point.
   Arrange the numbers from smallest to largest.

4) If any of the digits are the same, move onto the digits after the decimal point.
   Arrange the numbers from smallest to largest.

Step 1:	Step 2:	Step 3:	Step 4:
1.7	1.700	0.750	0.078
0.75	0.750	0.758	0.750
0.758	0.758	0.078	0.758
0.078	0.078	1.700	1.700

The order is: **0.078, 0.75, 0.758, 1.7**

Practice Questions

1) Which number is larger?

   a)   1.456 or 1.556   b)   3.145 or 3.142   c)   4.658 or 46.58

   ....

2) Which number is smaller?

   a)   0.658 or 0.75   b)   15.604 or 15.640   c)   6.431 or 6.481

   ....

3) Put these weights in order of size from smallest to largest:
   0.63 kg, 6.006 kg, 0.6 kg, 0.603 kg

   ....

Adding and Subtracting Decimals

You can add and subtract decimals using a calculator — just remember to type
the decimal point into the calculator.

EXAMPLE 1:

Sarah wants to know how much she spends on lunch on a weekend.

On Saturday, she spent £2.75 on a sandwich and £1.32 on a cup of coffee.
On Sunday, she spent £1.60 on some soup.

How much did she spend in total on lunch this weekend?

Answer: Add together everything Sarah has spent.

2.75 + 1.32 + 1.60 = **£5.67**

[2][.][7][5][+][1][.][3][2][+][1][.][6][0][=]   5.67

EXAMPLE 2:

Paul has been on a diet and lost some weight. He used to weigh 93.7 kg.
He now weighs 88.4 kg. How much weight has he lost?

Answer: Take away what Paul weighs now from what he used to weigh.

93.7 − 88.4 = **5.3 kg**

Adding Decimals Without a Calculator

1) Write the numbers one above the other, lining up the decimal points.

2) Make sure all the numbers have the same number of decimal places by filling in extra zeros at the ends where needed.

3) Write a decimal point on your answer line, lined up with the decimal points in the numbers above.

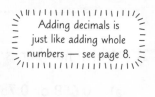

Adding decimals is just like adding whole numbers — see page 8.

4) Add up the columns from right to left. Remember to carry digits to the next column when you get a total that is greater than 9.

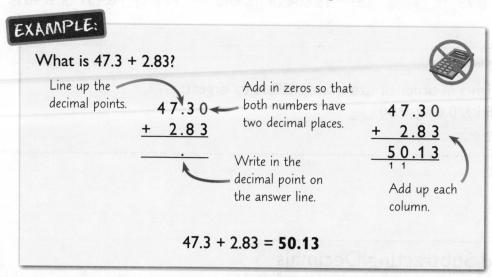

EXAMPLE:

What is 47.3 + 2.83?

Line up the decimal points.

Add in zeros so that both numbers have two decimal places.

```
  4 7 . 3 0
+     2 . 8 3
```

Write in the decimal point on the answer line.

```
  4 7 . 3 0
+     2 . 8 3
  5 0 . 1 3
    1   1
```

Add up each column.

47.3 + 2.83 = **50.13**

Subtracting Decimals Without a Calculator

1) Write the numbers one above the other, with the biggest number on top, lining up the decimal points.

2) Make sure all the numbers have the same number of decimal places by filling in extra zeros at the ends where needed.

3) Write a decimal point on your answer line, lined up with the decimal points in the numbers above.

Subtracting decimals is just like subtracting whole numbers — see page 9.

4) Subtract the numbers in each column from right to left. Remember to borrow when the top digit is smaller than the bottom one.

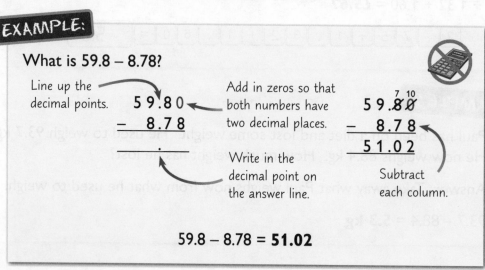

EXAMPLE:

What is 59.8 – 8.78?

Line up the decimal points.

Add in zeros so that both numbers have two decimal places.

```
  5 9 . 8 0
−     8 . 7 8
```

Write in the decimal point on the answer line.

```
        7  10
  5 9 . 8̸ 0̸
−     8 . 7 8
  5 1 . 0 2
```

Subtract each column.

59.8 – 8.78 = **51.02**

Practice Questions

1) What is 11.32 + 0.82?

....

2) Jenna wants to know how much her baby daughter Lucy has grown.
Lucy was 48.75 cm long. She is now 51.6 cm long. By how much has Lucy grown?

....

3) Matt goes to the garden centre. If he spends £20 or more he will get a free watering can.
Matt spends £12.99 on a trowel, £4.95 on flower pots and £1.62 on a bag of seeds.

Will Matt be given a free watering can? Explain your answer.

....

....

Multiplying and Dividing Decimals

You can multiply and divide decimals in exactly the same way as whole numbers.

EXAMPLE 1:

Lizzie makes 1.2 kg of fudge. She wants to give an equal amount to three friends. How much fudge should each friend get? Give your answer in kg.

Answer: Divide the amount of fudge by the number of friends.

$$1.2 \div 3 = \textbf{0.4 kg}$$

EXAMPLE 2:

Chris is working out his petrol expenses. He has driven 63.5 miles on business this week. He is allowed to claim £0.30 per mile for petrol.

How much money can Chris claim for petrol this week?
Give your answer in pounds (£).

Answer: Multiply the number of miles by the cost per mile.

$$63.5 \times 0.30 = \textbf{£19.05}$$

Multiplying Decimals Without a Calculator

1) Start by ignoring the decimal points.
   Just do the multiplication with whole numbers.

See page 10 for how to multiply numbers together.

2) Count the total number of digits after the decimal points in the original numbers.

3) Write in the decimal point to make the answer have the same number of decimal places. For example, 1.2 has 1 digit after the decimal point and 2.45 has 2 digits after the decimal point, so 1.2 × 2.45 would have 1 + 2 = 3 digits after the decimal point.

EXAMPLE:

What is 1.06 × 1.8?

1) Ignoring the decimal points, do the whole number multiplication.

$$106 \times 18 = 1908$$

This multiplication is shown on page 10.

2) Count the number of digits after the decimal points.

   $1.\underset{2}{06} \times 1.\underset{1}{8}$ has 2 + 1 = 3 digits after the decimal point.
   So the answer will have 3 digits after the decimal point too.

3) Write in the decimal point to give a number with the correct number of decimal places:

   $1.06 \times 1.8 = \textbf{1.}\underset{3}{\textbf{908}}$

Dividing a Decimal by a Whole Number Without a Calculator

1) When dividing a decimal by a whole number, set the question up as you would for whole numbers.

See page 11 for how to divide numbers.

2) Write a decimal point on the answer line directly above the decimal point in the number you're dividing.

3) Then do the division using the same method as for whole numbers.

EXAMPLE:

A zoo has a giraffe that is 5.94 m tall and an elephant that is half as tall as the giraffe. How tall is the elephant?

Answer: You need to divide 5.94 by 2.

1) Write the division with the number you're dividing by on the left.

$$2 \overline{)5.94}$$

2) Put a decimal point right above the one in the number you're dividing.

$$2 \overline{)5.94}$$

3) Use the same method to do the division as for whole numbers.

$$\begin{array}{r} 2.97 \\ 2 \overline{)5.{}^1 9 {}^1 4} \end{array}$$

So the elephant is **2.97 m** tall.

Dividing a Number by a Decimal Without a Calculator

1) Write the division as a fraction. The number you're dividing by goes on the bottom.

2) Multiply the top and bottom by 10, 100 or 1000 to remove
   any decimals and leave whole numbers on the top and bottom.
   You must multiply the top and the bottom by the same thing.

   Multiplying by 10, 100 and 1000 is covered on page 14.

3) Do the whole number division to give the answer.

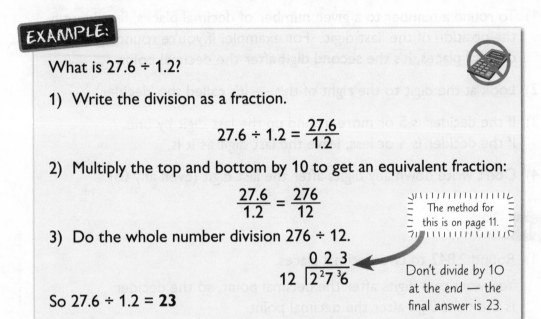

EXAMPLE:

What is 27.6 ÷ 1.2?

1) Write the division as a fraction.

$$27.6 \div 1.2 = \frac{27.6}{1.2}$$

2) Multiply the top and bottom by 10 to get an equivalent fraction:

$$\frac{27.6}{1.2} = \frac{276}{12}$$

The method for this is on page 11.

3) Do the whole number division 276 ÷ 12.

$$12 \overline{\smash)2^27^36} \quad 0\ 2\ 3$$

Don't divide by 10 at the end — the final answer is 23.

So 27.6 ÷ 1.2 = **23**

Practice Questions

1) What is 1.6 × 2.67?

....

....

2) Calculate 46.8 ÷ 3.

....

....

3) Brian has a 12.6 m roll of wallpaper. He needs to cut it into strips that are 2.4 m long.

   a) How many whole strips can he make?

   ....

   He is also painting the ceiling. He has bought 3 tins of paint. Each tin of paint
   covers 4.75 square metres. His ceiling has an area of 15 square metres.

   b) Does Brian have enough paint?

   ....

Rounding

Rounding off Decimals

You can sometimes get an answer with lots of numbers after the decimal point. Instead of writing down the whole thing, you can shorten the answer by rounding.

1) To round a number to a given number of decimal places, first identify the position of the 'last digit'. For example, if you're rounding to 2 decimal places, it's the second digit after the decimal point.

2) Look at the digit to the right of this — it's called the 'decider'.

3) If the decider is 5 or more, round up the last digit by one. If the decider is 4 or less, leave the last digit as it is.

4) Don't write down any digits after the last digit (even any 0s).

EXAMPLES:

1) Round 2.842 to two decimal places.

   You want two digits after the decimal point, so the decider is the third digit after the decimal point.

   Last digit   2.8 4 2   Decider
               $\quad_1\ _2\ _3$

   The decider is 2, which is less than 5, so leave the last digit as it is. So the answer is **2.84**.

2) Round 20.8537 to one decimal place.

   You want one digit after the decimal point, so the decider is the second digit after the decimal point.

   Last digit   20.8 5 3 7   Decider
              $\quad\ _1\ _2\ _3\ _4$

   The decider is 5, so you need to round the last digit up by one. So the answer is **20.9**.

If the last digit is a 9 and you have to round it up, you would need to round 9 to 10. This means the last digit becomes 0 and you add 1 to the digit to the left of the last digit.

EXAMPLE:

Round 5.398 to two decimal places.

Last
digit
$\qquad$ 5.398 $\rightarrow$ 5.3$\overset{4\ 0}{\cancel{9}}$ $\rightarrow$ **5.40**

Decider — round up

The question asks for two decimal places, so you need to include the 0 at the end — it's not just 5.4.

Rounding to the Nearest Whole Number

1) You can round to the nearest whole number in a similar way.

2) This time, the decider is the first digit after the decimal point.

EXAMPLES:

> 1) Round 9.318 to the nearest whole number.
>
>   The decider is the first digit after the decimal point, so it's 3.
>   This is less than 5, so the answer is **9**.
>
> 2) Round 25.65 to the nearest whole number.
>
>   The decider is 6, so you need to round up.
>   So the answer is **26**.

Practice Questions

1) Round each of these numbers to two decimal places.

   a) 3.4836

   b) 20.4569

   ....

   ....

2) Round each of these numbers to one decimal place.

   a) 0.925

   b) 6.555

   ....

   ....

3) Round these numbers to the given number of decimal places.

   a) 45.796 to 2 decimal places

   b) 2.971 to 1 decimal place

   ....

   ....

4) Round each of these numbers to the nearest whole number.

   a) 51.0684

   b) 5.847

   ....

   ....

5) The highest temperature recorded in a town is 35.3 °C.
   Round this temperature to the nearest degree.

   ....

Use Rounding to Estimate the Answers to Calculations

1) You can check your answers in tricky calculations by estimating.

2) Round each number in the calculation so that all the digits apart from the first one are zero (for example to the nearest whole number or the nearest 10). Then do the calculation using the rounded numbers.

3) If your estimate is close to the answer you got, then your answer is probably correct.

> **EXAMPLE:**
>
> Ronan calculates that 42 + 39.99 + 21.7 = 130.69.
> Use estimation to check his answer. Do you think it's correct?
>
> Round each of the numbers in the calculation to the nearest 10:
>
> $\quad$ 42 → 40 $\qquad$ 39.99 → 40 $\qquad$ 21.7 → 20
>
> Then do the calculation with your rounded numbers:
>
> $\quad$ 40 + 40 + 20 = 100
>
> 100 is quite far away from Ronan's answer, so he is probably **wrong**.

Practice Questions

1) Use rounding to estimate the answers to these calculations:

a) 31 + 98 + 1001

....

b) 41 × 2.1

....

c) (199 − 59) ÷ 6.9

....

2) The box on the right shows the amount Jacqueline spends each month on bills. Estimate the total amount that Jacqueline spends each month on bills.

Gas	£30.10
Electricity	£29.99
Internet	£10.05
Water	£19.80
Council tax	£104.43

....

....

Percentages

Understanding Percentages

1) 'Per cent' means 'out of 100'.

2) % is a short way of writing 'per cent'.

3) 20% means twenty per cent. This is the same as 20 out of 100.
   100% represents the whole amount of something.

4) You can write any percentage as a fraction.

$$20\% = \frac{20}{100}$$

Put the percentage on the top of the fraction.

Put 100 on the bottom of the fraction.

There's more on fractions on page 19.

Calculating Percentages

1) Sometimes, you might need to calculate the 'percentage of' something.

2) In these cases, 'of' means 'times' (multiply).

EXAMPLE 1:

What is 20% of 60?

1) Write it down: 20% of 60

2) Turn it into maths: $\frac{20}{100} \times 60$

3) Type it into your calculator: 20 ÷ 100 × 60 = **12**

EXAMPLE 2:

40 people are booked onto a trip. On the day, 15% of the people don't turn up. How many people don't turn up for the trip?

Answer: You just need to find 15% of 40.

1) Write it down: 15% of 40

2) Turn it into maths: $\frac{15}{100} \times 40$

3) Type it into your calculator: 15 ÷ 100 × 40 = **6**

Practice Questions

1) Write 35% as a fraction.

   ....

2) What is 15% of 50?

   ....

3) Kim teaches an aerobics class. The class has twenty members.

   a)   80% of the class are women. How many members of the class are women?

   ....

   b)   65% of the class are over 30. How many members of the class are over 30?

   ....

Finding 50% Without a Calculator

50% is just half of 100%. So to find 50%, divide 100% by 2.

50% of 600 = 600 ÷ 2 = 300	50% of 50 = 50 ÷ 2 = 25
50% of 18 = 18 ÷ 2 = 9	50% of 840 = 840 ÷ 2 = 420

Finding 10% and 5% Without a Calculator

1) You can write 10% as $\frac{10}{100}$. Using equivalent fractions, $\frac{10}{100} = \frac{1}{10}$.

2) To find $\frac{1}{10}$, you divide by 10. So to find 10%, just divide by 10.

> See page 21 for more on equivalent fractions and page 15 for how to divide by 10.

10% of 600 = 600 ÷ 10 = 60	10% of 50 = 50 ÷ 10 = 5
10% of 18 = 18 ÷ 10 = 1.8	10% of 840 = 840 ÷ 10 = 84

3) 5% is just half of 10%. So to find 5%, first find 10% (as above) and then divide by 2.

5% of 600 = 60 ÷ 2 = 30	5% of 50 = 5 ÷ 2 = 2.5
5% of 18 = 1.8 ÷ 2 = 0.9	5% of 840 = 84 ÷ 2 = 42

Finding Other Percentages Without a Calculator

You can find other percentages by breaking them down into lots of 5%, 10% and 50%.
For example, 30% = 10% + 10% + 10% (or 3 × 10%) and 15% = 10% + 5%.

EXAMPLE:

What is 65% of 400?

1) First, find 50%, 10% and 5% of 400.

$$50\% \text{ of } 400 = 400 \div 2 = 200$$
$$10\% = 400 \div 10 = 40$$
$$5\% = 40 \div 2 = 20$$

2) Then add together 50%, 10% and 5% to get 65%.

$$65\% = 50\% + 10\% + 5\% = 200 + 40 + 20 = \mathbf{260}$$

For bigger percentages you could subtract instead.
For example, 90% = 100% − 10% (100% is the original number).

Practice Questions

1) Find 10% of 360.

....

2) Find 5% of 260.

....

3) There are 160 dresses in stock at a shop at the start of the day. 10% of the dresses are
sold by the end of the day. How many dresses did the shop sell during the day?

....

4) What is 75% of 240?

....

5) A test is marked out of 110. A student gets 80% of the test right.
How many marks did they get?

....

Calculating Percentage Increase and Decrease

1) To find a percentage increase, you need to find the 'percentage of' first.
Then you add it on to the original number.

2) To find a percentage decrease, you need to find the 'percentage of' first.
Then you take it away from the original number.

EXAMPLE:

Last year, 760 people went to a festival. This year, there was
a 5% increase in the number of people who went to the festival.
How many people went to the festival this year?

1) Find 5% of 760: $\frac{5}{100} \times 760 = 5 \div 100 \times 760 = 38$

2) Add this on to 760: 760 + 38 = **798 people**

Practice Questions

1) Last year, Steve drove 3540 miles. This year, Steve has driven 15% further.

   a) How many more miles did Steve drive this year?

   ....

   b) How many miles in total has Steve driven this year?

   ....

2) Anja is buying material. She measures out 3 m. The shop owner gives
her an extra 10% for free. How much material does Anja end up with?

   ....

   ....

3) David runs a 10 km race and beats his best time by 5%. His best time
was 60 minutes. How long did he take to complete the race?

   ....

Fractions, Decimals and Percentages

These Fractions, Decimals and Percentages Are All the Same

The following fractions, decimals and percentages all mean the same thing.
They're really common, so it's a good idea to learn them.

$\frac{1}{2}$ is the same as 0.5, which is the same as 50%.

$\frac{1}{4}$ is the same as 0.25, which is the same as 25%.

$\frac{3}{4}$ is the same as 0.75, which is the same as 75%.

$\frac{1}{5}$ is the same as 0.2, which is the same as 20%.

$\frac{1}{10}$ is the same as 0.1, which is the same as 10%.

$\frac{1}{1}$ is the same as 1, which is the same as 100%.

You Can Change Fractions into Decimals

1) To convert a fraction into a decimal, you should:

Divide the top number in the fraction by the bottom number.

EXAMPLE 1:

What is $\frac{4}{5}$ as a decimal?

Divide 4 by 5:   4 ÷ 5 = **0.8**

EXAMPLE 2:

What is $\frac{9}{10}$ as a decimal?

Divide 9 by 10:   9 ÷ 10 = **0.9**

2) Use this method for trickier fractions when you don't have a calculator:

1) Multiply the top number by 10.

2) Divide by the bottom number.

3) Divide the result by 10.

EXAMPLE 3:

What is $\frac{4}{5}$ as a decimal?

1) Multiply the top number by 10:   4 × 10 = 40

2) Divide by the bottom number: 40 ÷ 5 = 8

3) Divide by 10:   8 ÷ 10 = **0.8**

You Can Change Fractions into Percentages

To change a fraction into a percentage, you should:

1) Multiply the fraction by 100.   2) Add a per cent (%) sign.

EXAMPLE:

Last year, Frank's horse won 3 of the 10 races it entered. What percentage of its races did Frank's horse win?

Frank's horse won 3 out of 10 races, so the fraction is $\frac{3}{10}$.

1) Multiply the fraction by 100.

$$\frac{3}{10} = 3 \div 10 = 0.3 \text{ and } 0.3 \times 100 = 30$$

2) Add a % sign: **30%**

See pages 14-15 for more on multiplying and dividing by 10 and 100.

Practice Questions

1) What is $\frac{1}{20}$ as a decimal?

....

2) A sale is offering a discount of 25%. What is this as a fraction?

....

3) Nadiya buys $1\frac{1}{4}$ kg of cheese. Write $1\frac{1}{4}$ kg as a decimal.

....

4) What is $\frac{3}{5}$ as:

a) a decimal? ....   b) a percentage? ....

5) Cumbria County Council sends out a survey. 40 out of 50 people respond.

a) What percentage is this?

....

b) Lancashire County Council sends out a similar survey. 7 out of 10 people respond. Which council has a higher percentage of people responding to their survey?

....

Comparing Fractions, Percentages and Decimals

You need to be able to compare fractions, percentages and decimals.

EXAMPLE 1:

Which is greater, 0.5 or $\frac{6}{10}$?

You need to work out what $\frac{6}{10}$ is as a decimal.

To convert $\frac{6}{10}$ to a decimal, divide 6 by 10: $6 \div 10 = 0.6$

0.6 is bigger than 0.5, so $\frac{6}{10}$ **is greater**.

EXAMPLE 2:

Tony is booking a holiday. The travel agent offers Tony two deals:

"All flights half price" OR "25% off all hotels"

The flights Tony wants to book normally cost £300.
The hotel he wants to book normally costs £400.

Which offer will save Tony the most money?

First work out how much money Tony will save on the flights: $\frac{1}{2} \times 300 = £150$

Then work out how much he'll save on the hotel:

25% of 400 = $\frac{25}{100} \times 400 = £100$

The **half price flights** offer will save Tony the most money.

Practice Questions

1) Which is greater, 0.25 or $\frac{2}{10}$?

....

2) Leanne buys a new TV in the sale. It would normally cost £500, but she gets 20% off.
Tamal also buys a new TV in the sale. It would normally cost £540, but he gets a third off.

Who ends up paying less for a TV, Leanne or Tamal? Explain your answer.

....

....

....

Ratios

Ratios Compare One Part to Another Part

Ratios are a way of showing how many things of one type there are compared to another.

EXAMPLE:

Look at this pattern:

There are two white tiles and six blue tiles.
In other words, for every one white tile there are three blue tiles.

So the ratio of white tiles to blue tiles is 1:3.

The order the numbers are written in the ratio depends on the order of the words —
the ratio of white tiles to blue tiles is 1:3. The ratio of blue tiles to white tiles is 3:1.

Questions Involving Ratios

To answer a question involving ratios, you usually need to start by working out the value of one part. For example, the cost of one thing or the mass of one part.

You can then use this to answer the question.

EXAMPLE 1:

5 pints of milk cost £3.00. How much will 3 pints cost?

1) First, you need to find out how much 1 pint of milk costs.
   You know that 5 pints cost £3, so you need to divide £3 by 5.

   cost of 1 pint = 3 ÷ 5 = £0.60

2) To work out the cost of 3 pints, times your answer by 3.

   £0.60 × 3 = **£1.80**

EXAMPLE 2:

A drink is made from 1 part cordial and 3 parts water.
800 ml of the drink is made. How much cordial is used?

The ratio of cordial to water is 1:3.

1) First, you need to work out how many parts there are in total.
   There's 1 part cordial and 3 parts water, so in total there are:

   1 + 3 = 4 parts

2) The drink contains 1 part cordial. To work out how many ml are
   in 1 part, divide the total amount by the total number of parts:

   800 ÷ 4 = **200 ml**

EXAMPLE 3:

£5000 will be split between two people in the ratio 1:4.
How much money does each person get?

1) First, work out how many parts the £5000 will be split into in total.
   To do this, add up the numbers in the ratio.

$$1 + 4 = 5 \text{ parts}$$

2) To find out how much one part is worth, divide 5000 by 5: $5000 \div 5 = 1000$

3) The first person in the ratio gets one part, so they get **£1000**.

4) The second person in the ratio gets four parts.
   To work out how much money they get, times the value of one part by 4:

$$1000 \times 4 = \textbf{£4000}$$

Practice Questions

1) Helen is making orange drink. She mixes 4 parts water to 1 part squash.

   a) What is the ratio of squash to water in Helen's drink?

   ....

   b) Helen wants to make 500 ml of orange drink. How much squash does she need?

   ....

   ....

2) Jake is tiling his bathroom floor. He uses three green tiles for every white tile.
   Jake uses twenty-four tiles in total. How many of them are green?

   ....

   ....

3) £3000 will be split between two people in the ratio 2:1. How much does each person get?

   ....

   ....

   ....

Working Out Total Amounts

1) You can use ratios to work out total amounts.

2) The first step is to work out the total number of parts.

3) The second step is multiplying the total number of parts by the value of one part.

> **EXAMPLE:**
>
> A drink is made from cordial and water in the ratio $1:3$.
> 70 ml of cordial is used. How much drink is made?
>
> 1) Find the total number of parts for the drink.
>   To do this, add up the numbers in the ratio: $1 + 3 = 4$
>
> 2) Multiply the total number of parts by the amount
>   given for one part: $4 \times 70 = $ **280 ml**   One part of the drink = 70 ml

Use Proportions to Scale Up and Down

Things are in proportion when they increase or decrease together in the same ratio.
You can use proportions to scale things up and down — for example, in recipes.

> **EXAMPLE 1:**
>
> Jia is making brownies. The recipe says to use 6 pieces of
> chocolate for every 2 eggs. Jia is using 10 eggs.
>
> How many pieces of chocolate does she need?
>
> Answer: 10 eggs is five times as many eggs as in the recipe.
>
> So Jia will need five times as many pieces of chocolate to match.
>
> $6 \times 5 = $ **30 pieces** of chocolate

> **EXAMPLE 2:**
>
> Nick is making lasagne. His recipe says to use 1 tin of tomatoes
> for every 500 g of mince. The recipe serves 4 people.
>
> Nick wants to make lasagne for 12 people.
> How much mince will he need to use?
>
> Answer: 12 people is three times as many as the recipe serves.
> So Nick needs to make three times as much lasagne.
>
> $500 \times 3 = $ **1500 g** of mince

Practice Questions

1) Hannah is mixing concrete. She mixes 1 part cement to 3 parts sand.
She uses 12 kg of cement. How much concrete will she have in total?

....

....

2) Danny is thinning some paint. He mixes 1 part paint thinner to 4 parts paint.
He uses 150 ml of paint thinner. How much thinned paint will he end up with?

....

....

3) Neil is making fairy cakes. The recipe says to use 120 g of flour for every 2 eggs.
The recipe makes 12 cakes.

   a) Neil uses 240 g of flour. How many eggs does he need to use?

   ....

   ....

   b) How much flour would Neil need to make 36 cakes?

   ....

   ....

4) Gina and Clive inherit some money. The money is split in a 1:2 ratio, with Clive getting
the most money. Gina inherits £500. How much money did the pair inherit in total?

....

....

5) Marie is making some lemonade. She needs 2 lemons for every 300 ml of water.
Marie has 900 ml of water. She wants to make as much lemonade as possible.
How many lemons will she need?

....

....

Formulas in Words

A Formula is a Type of Rule

1) A formula is a rule for working out an amount.

2) Formulas can be written in words. Sometimes, it can be tricky to spot the formula.

EXAMPLE:

Sam is paid £7.25 per hour. How much does he earn in 8 hours?

You're told that: "Sam is paid £7.25 per hour." This is a formula.
You can use it to work out how much Sam earns in a given number of hours.

1) The calculation you need to do here is:

$$\text{Sam's pay} = 7.25 \times \text{number of hours}$$

2) You've been asked how much Sam earns in 8 hours, so put '8' into the calculation in place of 'number of hours':

$$7.25 \times 8 = £58$$

You can use the same formula to work out how much Sam earns for any number of hours.

Formulas Can Have More Than One Step

Some formulas have two steps in them. You need to be able to use two-step formulas.

EXAMPLE:

A ham takes 30 minutes per kilogram to cook, plus an extra 25 minutes. How long does a 1.5 kg ham take to cook?

The formula here is "30 minutes per kilogram, plus 25 minutes".

1) Work out the calculation you need to do:

Step 1: $30 \times$ number of kilograms

Step 2: $+ 25$

Cooking time = $(30 \times \text{number of kilograms}) + 25$ ←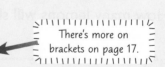
There's more on brackets on page 17.

2) Then just substitute the right numbers in.
In this case it's '1.5' in place of 'number of kilograms':

$$(30 \times 1.5) + 25 = \textbf{70 minutes}$$

Function Machines

Function machines can help you to use formulas with one or more than one step.

This function machine helps you to work out the cooking time for a ham. It gives you the cooking time of the ham in minutes.

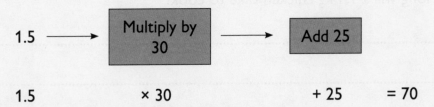

To work out how long a 1.5 kg ham takes to cook:

1) Put 1.5 into the function machine in place of 'Weight of ham in kg'.

2) Follow the rest of the steps.

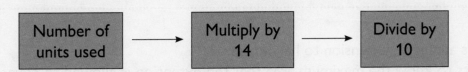

| 1.5 | × 30 | + 25 | = 70 |

So the ham takes **70 minutes** to cook.

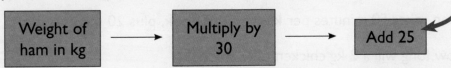

Carrie is checking her electricity bill.
To work out how much her electricity costs, she uses this rule:

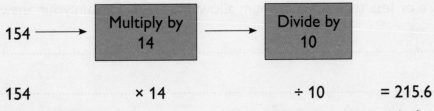

This gives her the cost of the electricity she has used in pounds (£).

Carrie has used 154 units of electricity so far this year.
How much will she be charged?

1) Put 154 into the function machine, in place of 'Number of units used'.

2) Follow the rest of the steps.

| 154 | × 14 | ÷ 10 | = 215.6 |

So Carrie will be charged **£215.60**.

Remember to write your answer in the correct money format (see page 47).

Practice Questions

1) Sanjay earns £7.60 per hour. How much will Sanjay earn in 7.5 hours?

   ....

   ....

2) A chicken takes 50 minutes per kilogram to cook, plus 20 minutes extra.

   a) How long will a 2 kg chicken take to cook?

   ....

   ....

   b) How long will a 1.4 kg chicken take to cook?

   ....

   ....

3) A company offers broadband for a one-off charge of £60 for the first 3 months,
   then £28 per month after that. How much will 12 months' broadband cost?

   ....

   ....

   ....

4) Charlie is adding an extension to her house.
   She wants to know the maximum area that the extension is allowed to cover.
   When she asks the council, they tell her to work it out using this rule:

   | Area of the house (in m²) | → | Divide by 4 | → | Add 10 |

   This will give Charlie the maximum area allowed for the extension in m^2.
   The area of Charlie's house is 90 m^2.

   Charlie wants to build an extension with an area of 32 m^2.
   Is this more or less than the maximum allowable area? Explain your answer.

   ....

   ....

   ....

Money

Working with Money

1) If you get a question on money, the units will probably be pounds (£) or pence (p).

2) You need to be able to switch between using pounds and using pence.

> To go from pounds to pence, multiply by 100.
>
> To go from pence to pounds, divide by 100.

Remember that
£1 = 100p.

3) You may get a question that uses pounds and pence.

4) If you do, you'll need to change the units so that they're all in pounds or all in pence.

EXAMPLE:

Cora buys a sandwich for £3.49, a cup of tea for £1.25 and a packet of crisps for 78p. How much does she need to pay in total?

1) Change the price of the crisps from pence to pounds.

$$78p \div 100 = £0.78$$

2) All the prices are in the same units now (£), so just add them up.

$$£0.78 + £3.49 + £1.25 = \textbf{£5.52}$$

You could also change the prices to pence, add them up, then change back to pounds.

5) If the question tells you what units to give your answer in then make sure you use those. If it doesn't, then it's a good idea to use pounds if the answer is more than 99p.

Practice Questions

1) a) What is £1.27 in pence?   b) What is 219p in pounds?

   ....

2) Bryn buys a turkey for £23, a loaf of bread for £1.99 and a pint of milk for 60p.
   How much does he spend in total?

   ....

   ....

   ....

Discounts or Increases Can Be Given as Percentages

Money questions often involve percentages.

For example, you might have to work out how much you save when you get a discount. You could also be asked the new price of something when it is increased or reduced by a certain percentage.

See pages 33-35 for more on working out percentages with or without your calculator.

EXAMPLE 1:

A jacket costs £42. How much money would be saved with a voucher for 15% off?

This question is just asking you to find 15% of £42.

1) Write it down:
$$15\% \text{ of } £42$$
$$\downarrow \quad \downarrow \quad \downarrow$$

2) Turn it into maths:
$$\frac{15}{100} \times 42$$

3) Type it into your calculator: $15 \div 100 \times 42 = $ **£6.30**

If you were asked to find the new amount after the discount, you would have to do £42 − £6.30 = £35.70.

EXAMPLE 2:

Priti goes on a cinema's website. She buys a film ticket for £5.80 and a gift card for £15. She is charged an online admin fee of 5%. How much does Priti pay in total?

1) Work out how much she pays before the admin fee: £5.80 + £15 = £20.80

2) Calculate 5% of £20.80: $\frac{5}{100} \times 20.80 = 5 \div 100 \times 20.80 = £1.04$

3) Add it on to £20.80: $20.80 + 1.04 = $ **£21.84**

You'd get the same answer by working out 5% of £5.80 and 5% of £15, then finding the total.

Interest

1) Interest is money that's added on to the value of something. It's given as a percentage.

2) For example, money saved in a bank account earns interest. Items that you buy on payment plans also cost more over time because interest is charged on them.

3) When working with interest, calculate the percentage of the amount and then add it to the original amount.

EXAMPLE:

Cara earns 5% interest on her savings. She has £50 in her account. How much money will she have once the interest has been added?

Find 5% of £50: $\frac{5}{100} \times 50 = 5 \div 100 \times 50 = £2.50$

Add this on to £50: $50 + 2.50 = $ **£52.50**

Practice Questions

1) A dress costs £35. How much could be saved with a voucher for 10% off?

   ....

2) Steve is buying a car. It would normally cost £3200, but today there is 20% off.
   What is the reduced price of the car?

   ....

3) A driving school offers individual lessons for £26 each. If you buy a block of 10 in one go,
   the school offers a discount of 15%. How much do you save by buying a block of 10?

   ....

   ....

4) A season ticket for a football club costs £330 if it is bought in April. If it is bought
   after April, it costs 25% more. How much does it cost if it is bought after April?

   ....

   ....

5) David borrows £5000 to start up a business. He pays back the loan in one year,
   plus 30% interest. How much money does David pay back in total?

   ....

   ....

6) Jemima is buying a lawn mower and a hedge trimmer. She wants to buy the items
   on a pay-monthly plan. The original prices in two shops are shown below.

Shop A		Shop B	
Lawn mower	£150	Lawn mower	£200
Hedge trimmer	£62	Hedge trimmer	£85

   Shop A charges 35% of the price of the items as interest. Shop B charges no interest.
   Work out which shop offers the items for the cheaper total cost.

   ....

   ....

   ....

Length

All Measures Have Units

1) Things that you measure have units. For example, metres (m) or grams (g).

2) They're really important. For example, you can't just say that a distance is 4 —
you need to know if it's 4 miles, 4 metres, 4 kilometres, etc.

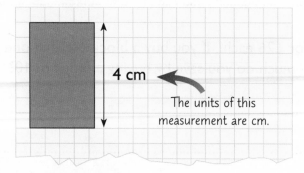

4 cm

The units of this
measurement are cm.

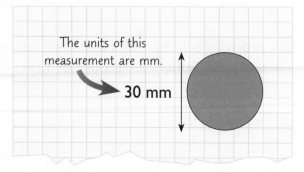

The units of this
measurement are mm.

30 mm

Units of Length

1) Length is how long something is. Some common units for length are millimetres (mm),
centimetres (cm), metres (m) and kilometres (km).

2) Here's how some of these units are related: ⟶

3) Sometimes you might need to change something
from one unit to another.

> 1 cm = 10 mm
> 1 m = 100 cm
> 1 km = 1000 m

4) To switch between mm, cm, m and km you can multiply or divide by 10, 100, or 1000.

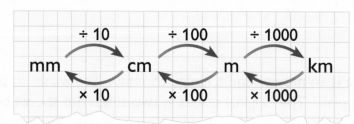

EXAMPLES:

1) How many millimetres are there in 20 cm?

Answer: You're going from cm to mm, so multiply by 10.

20 × 10 = **200 mm**

2) What is 3500 m in km?

Answer: You're going from m to km, so divide by 1000.

3500 ÷ 1000 = **3.5 km**

EXAMPLE:

Patrick is travelling from his home to his doctor's surgery. He walks 500 m to the bus stop, then gets on a bus. When he gets off the bus, he has to walk another 250 m.

In total, he travels 2.5 km.

How many kilometres does Patrick travel on the bus?

Answer: Change 500 m and 250 m into km: 500 m = 500 ÷ 1000 = 0.5 km
250 m = 250 ÷ 1000 = 0.25 km

So Patrick walks 0.5 + 0.25 = 0.75 km.

Then subtract this distance from 2.5 km to find how far he travels by bus:

2.5 − 0.75 = **1.75 km**

Practice Questions

1) How many cm are there in:

   a) 2 m?   b) 50 mm?   c) 1.5 m?

....

2) Ivy is planning to swim 5000 m for charity. How far is she planning to swim in kilometres?

....

3) Kirsten buys a dining table that is 1.7 m long. How long is the dining table in centimetres?

....

4) Roy is planning a journey for his sister and himself to the seaside. He needs to drive 5000 m from his house to his sister's house, then 170 km to the service station, then 60 km to the seaside. How far does he drive in total?

....

5) A wall in Lee's bedroom is 2.3 m wide. He buys a wardrobe that is 125 cm wide and a chest of drawers that is 90 cm wide. Will they both fit against the wall? Explain your answer.

....

....

Weight

Units of Weight

1) Weight is how heavy something is.
   Grams (g) and kilograms (kg) are common units for weight.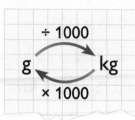

   1 kg = 1000 g

2) To change between g and kg, multiply or divide by 1000.

   ÷ 1000

   g ⟷ kg

   × 1000

EXAMPLE:

How many grams are there in 12 kg?

Answer: You're going from kg to g, so multiply by 1000.

12 × 1000 = **12 000 g**

Questions Involving Weight

You need to be able to solve problems involving weight.

EXAMPLE 1:

Roan has a bad back. His doctor told him not to lift more than 3000 g at a time.
Roan has bought one bag of flour, two bags of rice and four chocolate bars.

Flour
1.5 kg

Rice
500 g

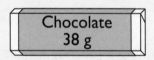
Chocolate
38 g

Can Roan carry all of his shopping back safely?

1) First you need everything in the same units,
   so change the weight of the flour into grams. ⟶ 1.5 kg × 1000 = 1500 g.

2) Next, work out the weight of the rice and chocolate.

   Don't forget — he's bought 2 bags of rice and 4 chocolate bars. ⟶ Rice: 500 g × 2 = 1000 g

   Chocolate: 38 g × 4 = 152 g

3) Then work out the total weight. ⟶ 1500 g + 1000 g + 152 g = 2652 g

   2652 g is less than 3000 g, so Roan **can** carry his shopping back safely.

Section Two — Measures, Shape and Space

EXAMPLE 2:

Arron has to take 5 boxes in a lift. The lift can carry 500 kg at a time. Each box weighs 100 kg. Arron weighs 90 kg. How many trips will he have to take?

1) Work out the total weight of the boxes. ➝ 5 × 100 kg = 500 kg

2) Work out the weight of the boxes plus Arron. ➝ 500 kg + 90 kg = 590 kg
   The lift can only carry 500 kg, so he'll need to take more than one trip.

3) Work out how many trips he needs to take.
   On the first trip he can take himself and 4 boxes: 4 × 100 kg + 90 kg = 490 kg
   On the second trip he can take himself and 1 box: 100 kg + 90 kg = 190 kg
   So it will take Arron **two trips** to take all of the boxes in the lift.

Practice Questions

1) Add up the following weights: 142 g, 263 g, 657 g, 4 g. Give your answer in kg.

....

2) Ewa is designing a child's chair. The chair needs to be able to carry a weight of 40 kg. How many 10 kg weights will Ewa need to test the strength of the chair?

....

3) Damon is a jockey. To compete in the local horse race, he and his equipment have to weigh less than 57 kg. Damon weighs 51 kg and his equipment weighs 5400 g. Can Damon take part in the race? Explain your answer.

....

....

4) Washing powder comes in three different sized boxes: 1 kg, 2.5 kg and 5 kg. Fin wants to buy exactly 14 kg of washing powder. What is the smallest number of boxes Fin can buy?

....

....

5) Samah has made 20 kg of chilli. There are 100 g of chilli in a portion. How many portions of chilli has Samah made?

....

Capacity

Volume and Capacity

Volume is the amount of 3D space something takes up.

Capacity is how much something will hold.

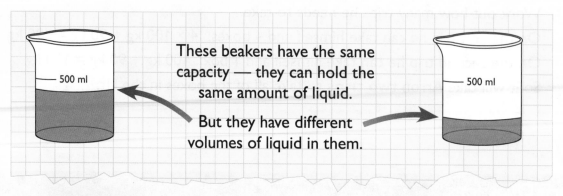

500 ml

These beakers have the same capacity — they can hold the same amount of liquid.

But they have different volumes of liquid in them.

500 ml

EXAMPLE:

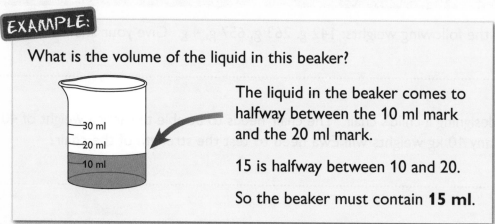

What is the volume of the liquid in this beaker?

30 ml
20 ml
10 ml

The liquid in the beaker comes to halfway between the 10 ml mark and the 20 ml mark.

15 is halfway between 10 and 20.

So the beaker must contain **15 ml**.

Units of Capacity

1) Common units of volume and capacity are millilitres (ml) and litres (L).

2) To change between ml and L, you can multiply or divide by 1000.

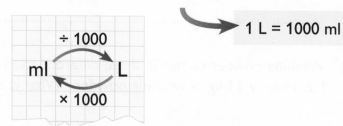

÷ 1000

ml   L

× 1000

1 L = 1000 ml

EXAMPLE:

How many millilitres are in 3 L?

Answer: You're going from L to ml, so multiply by 1000.

3 × 1000 = **3000 ml**

Questions Involving Capacity and Volume

You need to be able to solve problems involving capacity and volume.

EXAMPLE:

Andrew needs to measure out 15 ml of vanilla essence.
He has one 10 ml and one 2.5 ml measuring spoon.
How can Andrew accurately measure out his vanilla essence?

Answer: He can measure out 10 ml using the 10 ml measuring spoon.

Then he can measure out two lots of 2.5 ml
using the 2.5 ml measuring spoon.

10 ml + 2.5 ml + 2.5 ml = **15 ml**

You can't accurately measure out
5 ml using half of the 10 ml spoon,
so you have to use the 2.5 ml spoon.

Practice Questions

1) Imran needs to use 500 ml of wine to make a stew. His measuring jug only measures in L.
    How many litres of wine should he use?

    ....

2) Brianna wants to paint her dining room. She needs 10 L of paint.
    The paint only comes in 2 L tins. How many tins of paint does Brianna need to buy?

    ....

3) Jason pours the water shown in the containers on the right
    into his water bottle. The capacity of his water bottle is 1 L.
    How much space is left in his water bottle in ml?

    ....

    ....

4) Hayden is designing a new cocktail. He wants it to fit in a jug that holds 400 ml.
    The ingredients he uses are shown below.

    | 0.25 L orange juice   50 ml lemonade   50 ml cranberry juice   0.1 L lime juice |

    Will the cocktail fit into the jug? Explain your answer.

    ....

    ....

Time

Time Has Lots of Different Units

You need to be able to use lots of different units for time. You also need to be able to change between them. Here are how some of the units of time are related:

60 seconds = 1 minute	7 days = 1 week	10 years = 1 decade
60 minutes = 1 hour	365 days = 1 year	100 years = 1 century
24 hours = 1 day	12 months = 1 year	

15 minutes = a quarter of an hour

30 minutes = half an hour

45 minutes = three quarters of an hour

The 12-Hour Clock and the 24-Hour Clock

1) You can give the time using the 12-hour clock or the 24-hour clock.

2) The 24-hour clock goes from 00:00 (midnight) to 23:59 (one minute before the next midnight).

> 01:00 is 1 o'clock in the morning. 13:00 is 1 o'clock in the afternoon.

3) The 12-hour clock goes from 12:00 am (midnight) to 11:59 am (one minute before noon), and then from 12:00 pm (noon) till 11:59 pm (one minute before midnight).

> 3 am is 3 o'clock in the morning. 3 pm is 3 o'clock in the afternoon.

4) For times in the afternoon, you need to add 12 hours to go from the 12-hour clock to the 24-hour clock. You take away 12 hours to go from the 24-hour clock to the 12-hour clock.

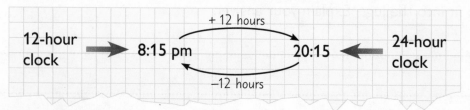

If the hours number is 12 or less, you don't need to subtract 12.

Practice Questions

1) How many minutes are there in 2 hours?

   ....

2) How many years are there in 3 decades?

   ....

3) How many minutes is 240 seconds?

   ....

4) Change the times below from the 24-hour clock to the 12-hour clock.

   a)   09:00   b)   16:45

   ....

5) Change the times below from the 12-hour clock to the 24-hour clock.

   a)   5:15 pm   b)   7:05 am

   ....

6) Chloe gets to a bus stop at 10:47 pm. The last bus leaves at 22:55. Has she missed it?

   ....

Working Out Lengths of Times

To work out how long something took, break it into stages.

EXAMPLE:

Claire caught a train at 8:30 am and arrived at 11:17 am.
How long was her journey?

8:30 am ——→ 9:00 am ——→ 11:00 am ——→ 11:17 am
       30 mins   2 hours   17 mins

Add up the hours and minutes separately: 2 hours
                                            30 mins + 17 mins = 47 mins

So the journey took **2 hours and 47 mins**.

Working Out Times

1) You may need to work out what time something will happen.
   For example, what time something will start or finish, or when to meet someone.

2) The best way to do this is to split the time into chunks.

EXAMPLE 1:

Melissa is visiting her sister. The drive takes 3 hours and 30 minutes.

She'll stop at a service station for a 20 minute break.

If she leaves at 6:00 pm, what time should she arrive?

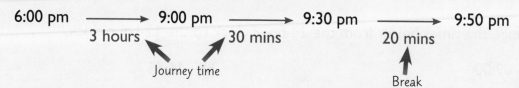

6:00 pm — 3 hours → 9:00 pm — 30 mins → 9:30 pm — 20 mins → 9:50 pm

Journey time

Break

She should arrive at **9:50 pm**.

It doesn't matter where in the journey she takes the break. The total time will be the same.

EXAMPLE 2:

Pranav is meeting a friend for lunch.

He has a half hour meeting with his bank manager that starts at 11 am.

He has some shopping to do straight after the meeting, which will take 40 minutes.

It will then take him 25 minutes to drive to the cafe for lunch.

What is the earliest time Pranav should arrange to meet his friend?

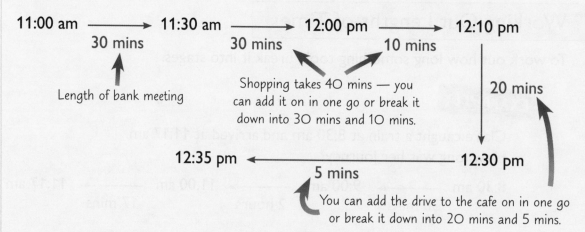

11:00 am — 30 mins → 11:30 am — 30 mins → 12:00 pm — 10 mins → 12:10 pm

Length of bank meeting

Shopping takes 40 mins — you can add it on in one go or break it down into 30 mins and 10 mins.

20 mins

12:35 pm ← 5 mins — 12:30 pm

You can add the drive to the cafe on in one go or break it down into 20 mins and 5 mins.

So the earliest he can be at the cafe is **12:35 pm**.

To be on the safe side he might arrange to meet this friend sometime after 12:35 pm.

Practice Questions

1) A film starts at 7:15 pm and finishes at 9:30 pm. How long is the film?

....

....

2) Daj catches a bus at 8:55 am and gets off at 11:47 am. How long was the bus trip?

....

....

3) Chris is cooking dinner. His recipe says it will take 35 minutes to prepare and 45 minutes to cook. If he starts at 6:30 pm, what time will dinner be ready?

....

....

4) Mary arrives at a museum at 3:35 pm. The museum closes at 5:30 pm.
How long will Mary be able to spend in the museum?

....

....

5) Flavia has 6 wedding invitations to make. It takes her 20 minutes to make each invitation. She'll stop for half an hour to eat dinner. If she starts at 17:45, what time will she finish?

....

....

....

6) Ashley would like to go to the theatre to see a show that starts at 19:30.
He usually gets home from work at 5:45 pm.
He reckons it will take him half an hour to get ready, and 45 minutes to drive to the theatre.
He wants to leave himself a quarter of an hour to find a parking space.
Will he be able to get to the theatre in time?

....

....

....

Length and Perimeter

Finding the Perimeter of a Shape

The perimeter is the distance around the outside of a shape.

To find a perimeter, you add up the lengths of all the sides.

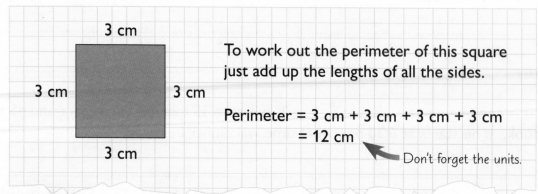

To work out the perimeter of this square just add up the lengths of all the sides.

Perimeter = 3 cm + 3 cm + 3 cm + 3 cm
= 12 cm

Don't forget the units.

EXAMPLE:

Find the perimeter of the shape below.

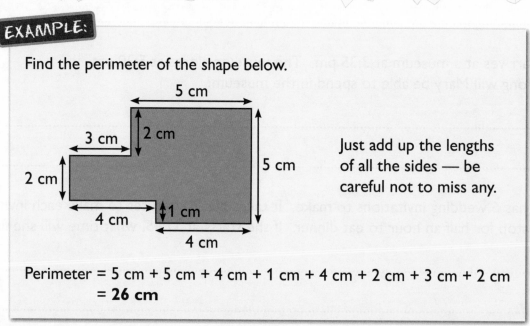

Just add up the lengths of all the sides — be careful not to miss any.

Perimeter = 5 cm + 5 cm + 4 cm + 1 cm + 4 cm + 2 cm + 3 cm + 2 cm
= **26 cm**

Practice Question

1) Find the perimeter of each of the following shapes.

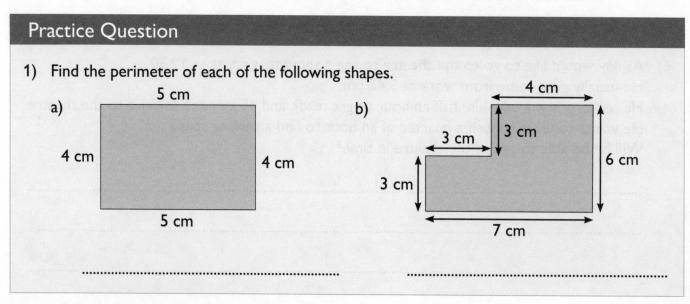

a)

b)

Working Out the Length of an Unknown Side

If you're only given the lengths of some of the sides, you'll have to work out the rest before you can calculate the perimeter. Sometimes this is fairly simple.

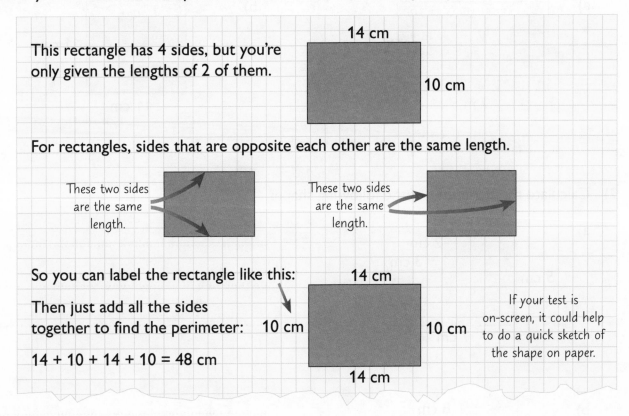

This rectangle has 4 sides, but you're only given the lengths of 2 of them.

14 cm

10 cm

For rectangles, sides that are opposite each other are the same length.

These two sides are the same length.

These two sides are the same length.

So you can label the rectangle like this:

Then just add all the sides together to find the perimeter:

14 + 10 + 14 + 10 = 48 cm

14 cm

10 cm   10 cm

14 cm

If your test is on-screen, it could help to do a quick sketch of the shape on paper.

It's a bit harder to find the lengths of unknown sides if you're not dealing with rectangles.

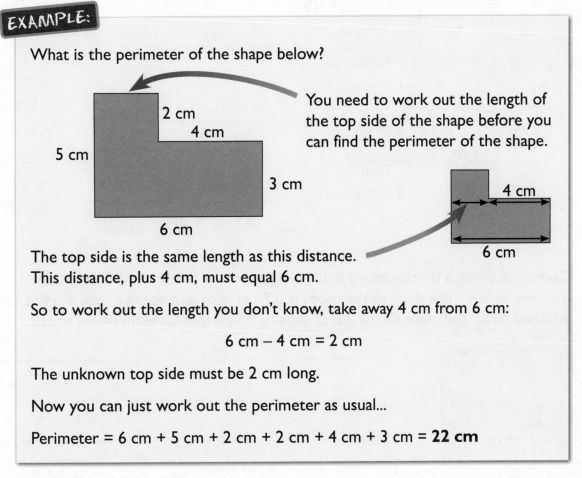

EXAMPLE:

What is the perimeter of the shape below?

2 cm
4 cm
5 cm
3 cm
6 cm

You need to work out the length of the top side of the shape before you can find the perimeter of the shape.

4 cm
6 cm

The top side is the same length as this distance. This distance, plus 4 cm, must equal 6 cm.

So to work out the length you don't know, take away 4 cm from 6 cm:

6 cm − 4 cm = 2 cm

The unknown top side must be 2 cm long.

Now you can just work out the perimeter as usual...

Perimeter = 6 cm + 5 cm + 2 cm + 2 cm + 4 cm + 3 cm = **22 cm**

Section Two — Measures, Shape and Space

Practice Questions

1) Calculate the perimeter of the following shapes.

   a)

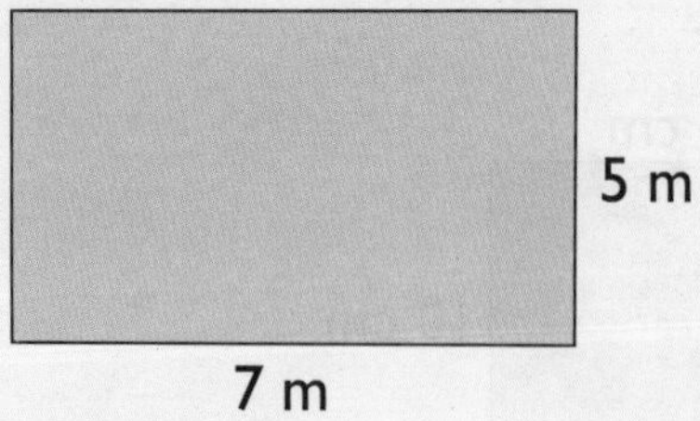

   5 m

   7 m

   b)

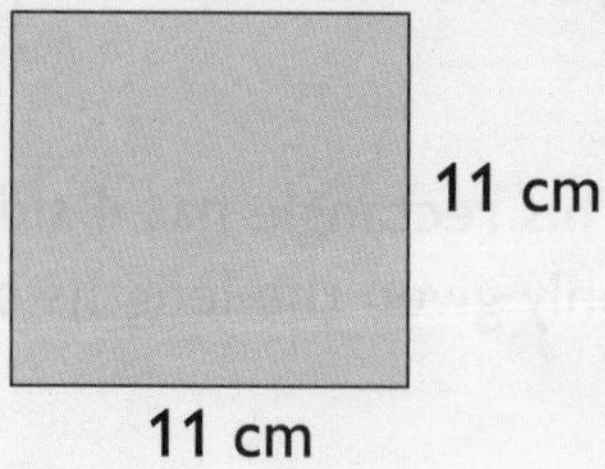

   11 cm

   11 cm

   ....

2) Calculate the perimeter of the following shapes.

   a)

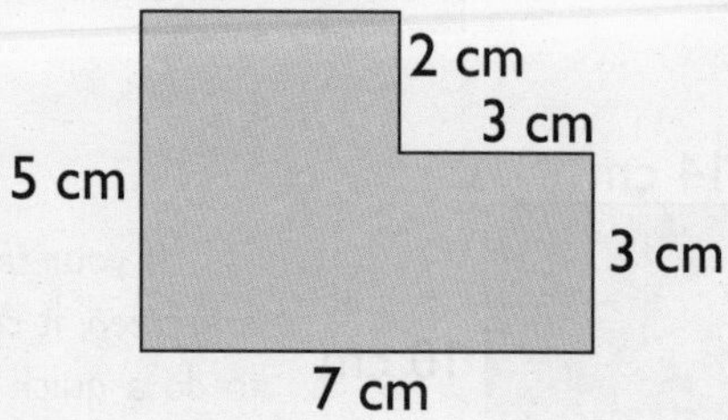

   2 cm
   3 cm
   5 cm
   3 cm
   7 cm

   ....
   ....
   ....

   b)

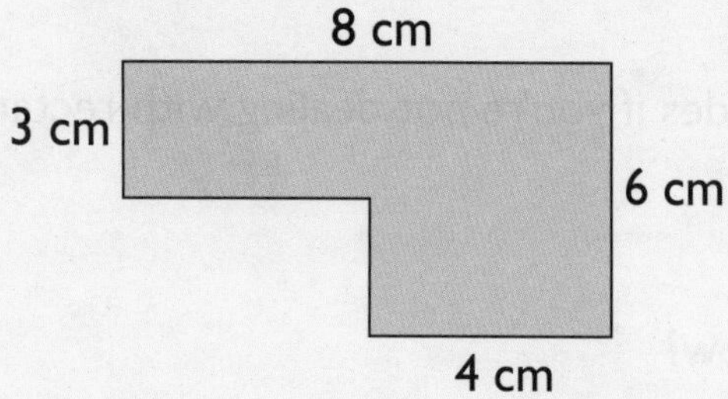

   8 cm
   3 cm
   6 cm
   4 cm

   ....
   ....
   ....

   c)

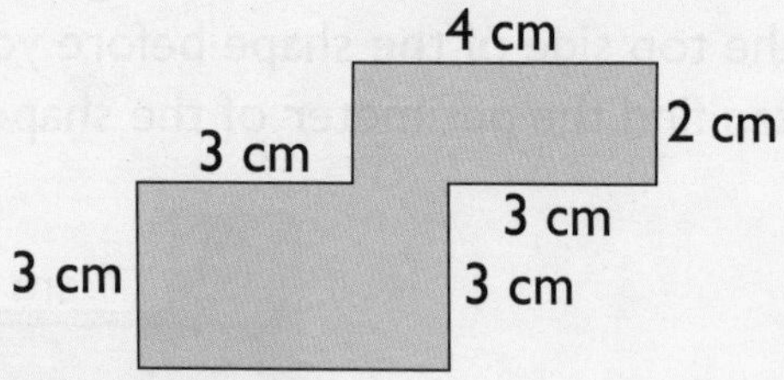

   4 cm
   3 cm
   2 cm
   3 cm
   3 cm
   3 cm

   ....
   ....
   ....

3) Quentin is fitting a skirting board in his living room. A sketch of his living room is shown below. The door to the room is 0.75 m wide and doesn't need skirting board attached to it. Calculate the length of skirting board that Quentin needs to buy.

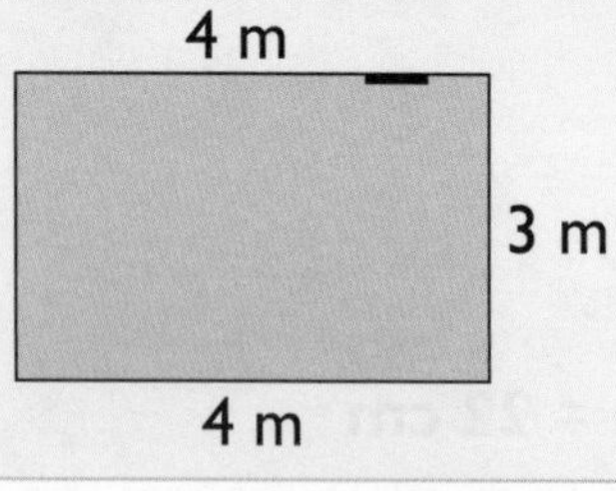

   4 m
   3 m
   4 m

   ....
   ....
   ....

Section Two — Measures, Shape and Space

Area

You Can Find the Area of Shapes by Counting...

1) Area is how much surface a shape covers.

2) If a shape's on a square grid, count how many squares it covers to find its area.

3) The squares may have sides one centimetre long.
   If so, each square is 1 centimetre squared. This is written as 1 cm².

EXAMPLE:

Find the area of the rectangle on the right.

There are 6 squares and each square has sides 1 cm long.

So the area of the shape is **6 cm²**.

cm² are the units.

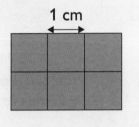

1 cm

4) If you're dealing with area, the units will be something squared.
   For example, cm², m², mm².

...or by Multiplying

1) You can work out the area of rectangles by multiplying.

2) You need to know the lengths of the sides, then just multiply them together.

EXAMPLE 1:

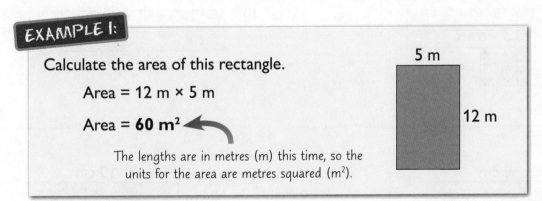

Calculate the area of this rectangle.

Area = 12 m × 5 m

Area = **60 m²**

The lengths are in metres (m) this time, so the
units for the area are metres squared (m²).

5 m

12 m

EXAMPLE 2:

Sophie is buying a new carpet for her dining room. The room is 6 m long
and 5 m wide. What area of carpet does she need to buy?

Answer: 6 m × 5 m = **30 m²**

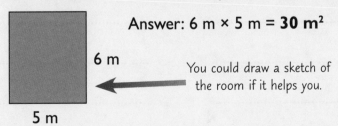

6 m

You could draw a sketch of
the room if it helps you.

5 m

Sometimes You Need to Split Shapes Up to Find the Area

It's a bit trickier to find the area of a shape that isn't a rectangle...

...but you can sometimes do it by splitting the shape up into rectangles.

EXAMPLE:

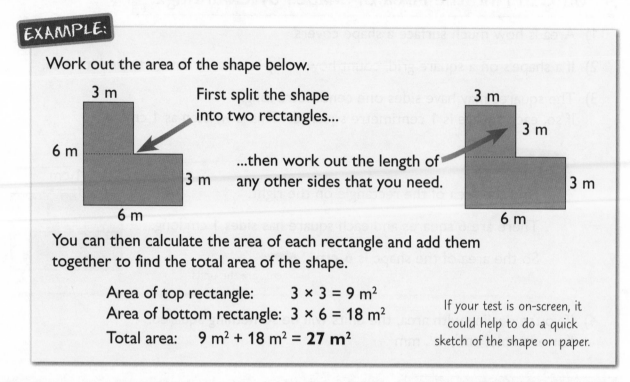

Work out the area of the shape below.

First split the shape into two rectangles...

...then work out the length of any other sides that you need.

You can then calculate the area of each rectangle and add them together to find the total area of the shape.

Area of top rectangle: $3 \times 3 = 9$ m^2
Area of bottom rectangle: $3 \times 6 = 18$ m^2
Total area: 9 m^2 + 18 m^2 = **27 m^2**

If your test is on-screen, it could help to do a quick sketch of the shape on paper.

Practice Question

1) Find the area of each of the shapes below. Don't forget to give the units.

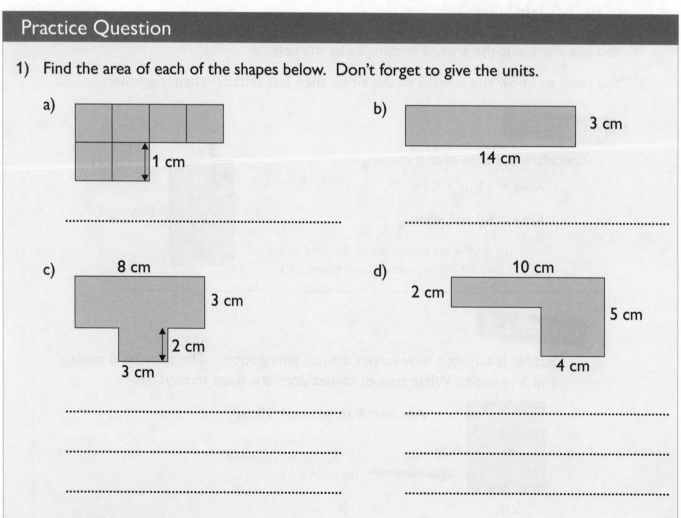

a)

1 cm

b)

3 cm

14 cm

c)

8 cm

3 cm

2 cm

3 cm

d)

10 cm

2 cm

5 cm

4 cm

Using Areas in Calculations

Sometimes you'll need to work out an area as part of a bigger calculation.

EXAMPLE 1:

Thomas is painting two walls, shown below. One wall is 7 m long and 3 m high, the other is 6.5 m long and 2 m high.

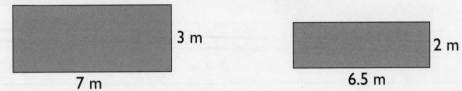

3 m

7 m

2 m

6.5 m

Each tin of paint will cover 12.5 m².
How many tins of paint does he need to buy?

First work out the area of each wall:

$$\text{Area of wall 1:} \quad 7 \text{ m} \times 3 \text{ m} = 21 \text{ m}^2$$
$$\text{Area of wall 2:} \quad 6.5 \text{ m} \times 2 \text{ m} = 13 \text{ m}^2$$

Then add them together to find the total area: 21 m² + 13 m² = 34 m²

Now work out how many tins of paint Thomas needs to cover this area.
To do this, divide the total area of the walls by the area that one tin will cover:

$$34 \text{ m}^2 \div 12.5 \text{ m}^2 = 2.72$$

Thomas can't buy 2.72 tins of paint, so he'll have to buy **3 tins**.

2 tins of paint wouldn't be enough, so you need to round up to 3.

EXAMPLE 2:

Christine is laying a patio in her backyard. The patio will be 4 m long and 3 m wide. The paving stones she'll use are 0.5 m long and 0.5 m wide.

How many paving stones does Christine need to buy?

You need to work out how many paving stones will fit into the area of the patio.
So first calculate the area of the patio: 3 × 4 = 12 m²

Then calculate the area of one paving stone: 0.5 × 0.5 = 0.25 m²

Now divide the area of the patio by the area of one paving stone:

$$12 \div 0.25 = \textbf{48 paving stones}$$

Practice Questions

1) A social club is raising money to put gravel on their car park and path. It will cost £4.20 for each m² that needs gravel. How much money does the social club need to raise?

car park   10 m

15 m

path   1.5 m

10 m

....

....

....

....

....

2) Emma is tiling a section of wall in her bathroom. The tiles are 10 cm by 10 cm. The section of wall she wants to tile is 120 cm across and 80 cm high.

Section of wall   80 cm

120 cm

Tile   10 cm

10 cm

How many tiles will Emma need?

....

....

....

....

Volume

You Can Calculate the Volume of a Shape by Counting...

1) Volume is how much 3D space something takes up.

2) Sometimes you can calculate the volume of a shape by counting cubes.

3) The units for volume will be something 'cubed'.
For example, cm³ (centimetres cubed) or m³ (metres cubed).

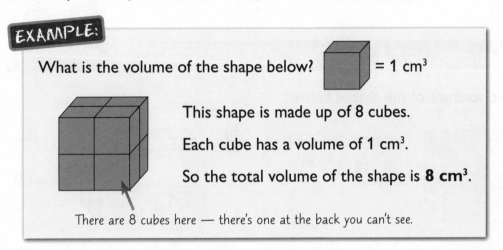

EXAMPLE:

What is the volume of the shape below?   = 1 cm³

This shape is made up of 8 cubes.

Each cube has a volume of 1 cm³.

So the total volume of the shape is **8 cm³**.

There are 8 cubes here — there's one at the back you can't see.

...or by Multiplying

1) You can work out the volume of shapes even if they aren't broken down into cubes.

2) For some shapes, you just need to know the length, the width and the height.
Then you just multiply them together.

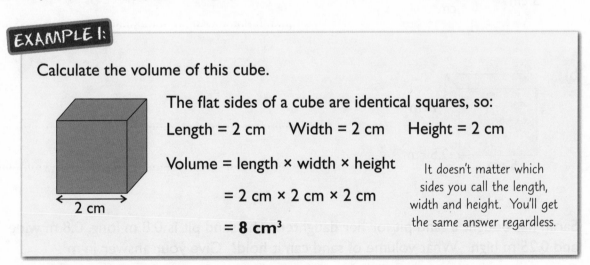

EXAMPLE 1:

Calculate the volume of this cube.

The flat sides of a cube are identical squares, so:

Length = 2 cm   Width = 2 cm   Height = 2 cm

Volume = length × width × height

= 2 cm × 2 cm × 2 cm

= **8 cm³**

It doesn't matter which sides you call the length, width and height. You'll get the same answer regardless.

2 cm

3) The units are cm³ in the example above, because you've multiplied three lots
of cm together. If the sides were measured in m, the units for volume would be m³.

EXAMPLE 2:

Calculate the volume of the cuboid on the right.

Length = 8 cm   Width = 2 cm   Height = 4 cm

Volume = length × width × height

= 8 cm × 2 cm × 4 cm

= **64 cm³**

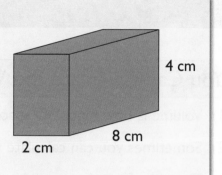

4 cm

8 cm

2 cm

Practice Questions

1) What are the volumes of the shapes below?

a)

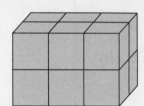

= 1 cm³

b)

= 1 cm³

....

....

2) Calculate the volumes of the shapes below.

a)

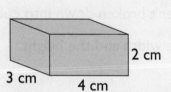

2 cm

3 cm   4 cm

....

....

b)

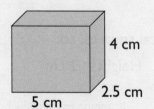

4 cm

2.5 cm

5 cm

....

....

3) Sarah has bought a sand pit for her daughter. The sand pit is 0.8 m long, 0.8 m wide and 0.25 m high. What volume of sand can it hold? Give your answer in m³.

....

....

2D Shapes

Some Shapes Have Lines of Symmetry

1) Shapes with a line of symmetry have two halves that are mirror images of each other.

2) You could fold a shape along this line and the sides would fold exactly together.

3) Some shapes have more than one line of symmetry.

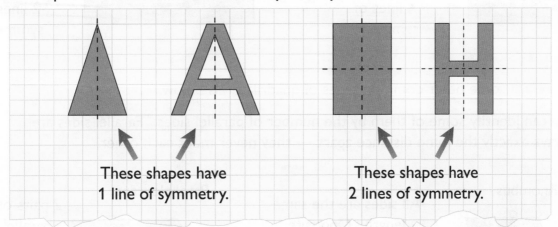

These shapes have 1 line of symmetry.

These shapes have 2 lines of symmetry.

4) Some shapes have no lines of symmetry.

Triangles

Triangles are 3-sided shapes.

They have different names depending on how many of their sides are the same length.

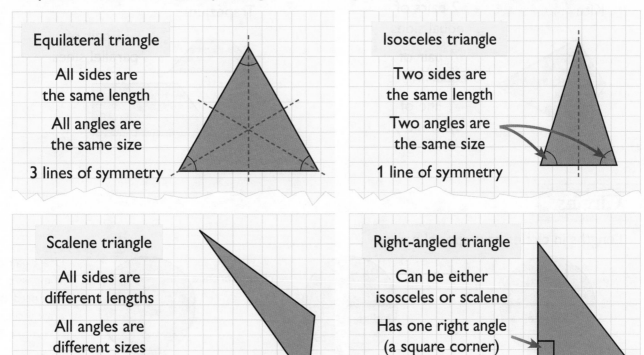

Equilateral triangle	Isosceles triangle
All sides are the same length	Two sides are the same length
All angles are the same size	Two angles are the same size
3 lines of symmetry	1 line of symmetry

Scalene triangle	Right-angled triangle
All sides are different lengths	Can be either isosceles or scalene
All angles are different sizes	Has one right angle (a square corner)
0 lines of symmetry	

See page 76 for more on angles.

Quadrilaterals

Quadrilaterals are 4-sided shapes. You've probably seen these ones before...

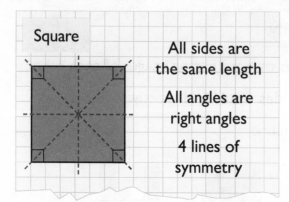

Square

All sides are the same length

All angles are right angles

4 lines of symmetry

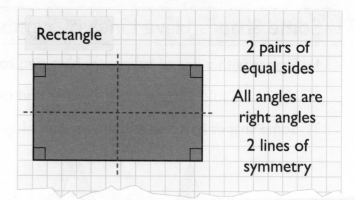

Rectangle

2 pairs of equal sides

All angles are right angles

2 lines of symmetry

There are some other common types of quadrilaterals below.

You can identify some of them by the number of parallel sides. Parallel sides are always exactly the same distance apart — they never meet each other.

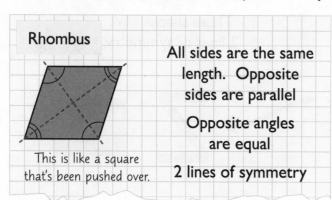

Rhombus

All sides are the same length. Opposite sides are parallel

Opposite angles are equal

This is like a square that's been pushed over.

2 lines of symmetry

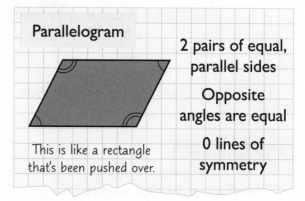

Parallelogram

2 pairs of equal, parallel sides

Opposite angles are equal

This is like a rectangle that's been pushed over.

0 lines of symmetry

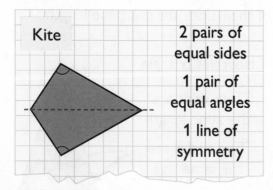

Kite

2 pairs of equal sides

1 pair of equal angles

1 line of symmetry

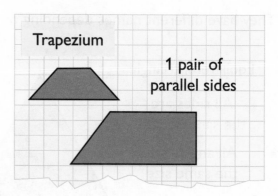

Trapezium

1 pair of parallel sides

Circles

Lengths in circles have special names:

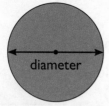

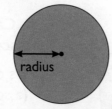

circumference

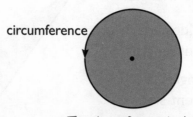

Use a pair of compasses if you need to draw a circle.

The diameter is the distance from one side to the other, passing through the centre.

The radius is the distance from the side to the centre. It's half of the diameter.

The circumference is the distance around the circle (so it's the perimeter).

Practice Questions

1) Draw the line (or lines) of symmetry on the shapes below:

a)

b)

c)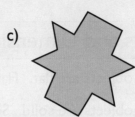

2) Complete the shapes below so they are symmetrical in the lines of symmetry shown.

a)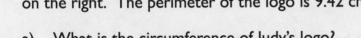

b)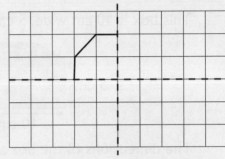

3) Judy is designing a logo in the shape of a circle, shown on the right. The perimeter of the logo is 9.42 cm.

1 cm

1 cm

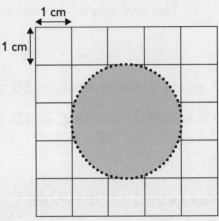

a) What is the circumference of Judy's logo?

....

b) What is the diameter of Judy's logo?

....

4) One of the angles in each of the shapes below has been highlighted.
Circle all the angles that are the same size as the highlighted one.

a)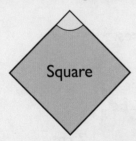
Square

b)
Kite

c)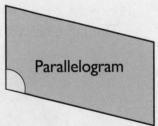
Parallelogram

5) Padma has bought a new rug in the shape of a trapezium.

- The rug has no lines of symmetry.
- Exactly two of the sides are each 2 m long.
- It has two right angles.

Draw a diagram on this grid of what the shape of the rug might look like.

1 m

1 m

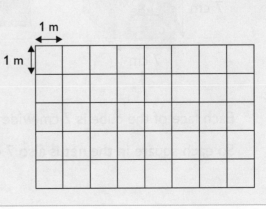

Nets, Plans and Elevations

Objects and Dimensions

1) Some objects are flat. Flat objects are called 2D objects.

2) Some objects are solid. Solid objects are called 3D objects.

3) The dimensions of an object tell you its size.

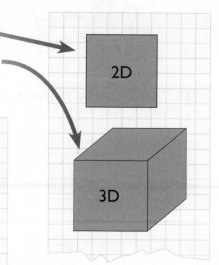

This box is 10 cm wide, 5 cm high and 6 cm deep.

5 cm

6 cm

10 cm

The dimensions of the box are 10 cm by 5 cm by 6 cm.
This can also be written as 10 cm × 5 cm × 6 cm.

2D means '2-dimensional', so 2D objects have 2 dimensions — e.g. width and height.

3D means '3-dimensional', so 3D objects have 3 dimensions — e.g. width, height and depth.

You Need to be Able to Draw Nets

A net is just a 3D shape folded out flat. You can use a net to help you make a 3D object.

The nets for cubes and boxes can usually been shown in the same basic shape.

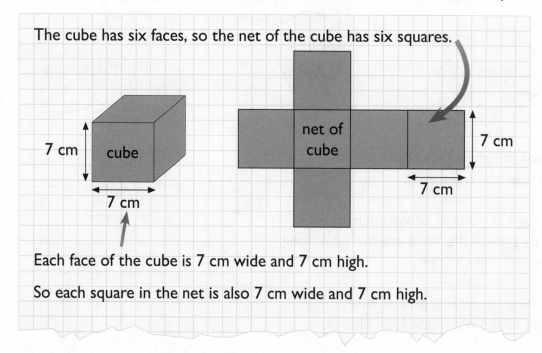

The cube has six faces, so the net of the cube has six squares.

7 cm

cube

7 cm

net of
cube

7 cm

7 cm

Each face of the cube is 7 cm wide and 7 cm high.

So each square in the net is also 7 cm wide and 7 cm high.

The Net of a Cuboid is Made Up of Rectangles

The faces of a cuboid are all rectangles, so the net will be made up from rectangles.

Look at the dimensions of the cuboid to work out the dimensions of the faces.

EXAMPLE:

Draw a net for the box below.

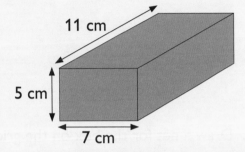

The box has 6 faces, so the net for the box will be made from 6 rectangles.

1) Draw the rectangle for the bottom of the box first.
   The diagram tells you it should be 11 cm long and 7 cm wide.

2) Next draw the rectangles for the sides of the box.
   The sides should be 11 cm long and 5 cm wide.

3) Now draw the rectangles for the front and back of the box.
   These should be 5 cm long and 7 cm wide.

4) Finally, draw the top of the box. The top of the box should be
   the same length and width as the bottom of the box.

You should end up with something like this:

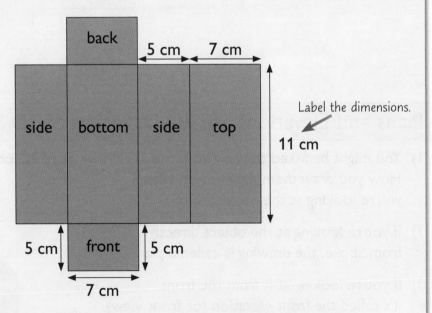

Tip: try to imagine your net being folded back up into the box.
If it works, there's a good chance you've got it right.

Practice Questions

1) A cube is shown below. Sketch a net for the cube in the space to the right of it. Your sketch does not need to be full size, but you should label the dimensions.

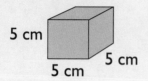

5 cm

5 cm

5 cm

2) Look at the box below. Draw a net for the box on the grid. Key: 1 square = 1 cm. Label the dimensions.

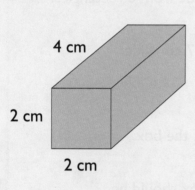

4 cm

2 cm

2 cm

Plans and Elevations are 2D Drawings of 3D Shapes

1) You might be asked to draw accurate 2D drawings of 3D objects. How you draw them depends on where you're looking at the object from.

2) If you're looking at the object directly from above, the drawing is called a plan.

3) If you're looking at it from the front, it's called the front elevation (or front view).

4) If you're looking at it from the side, it's called the side elevation (or side view).

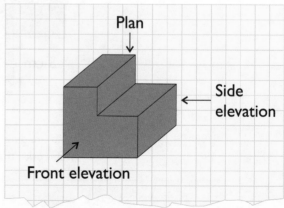

Plan

Side elevation

Front elevation

EXAMPLE:

Draw the plan, front elevation and side elevation of the object on the right.

1 square on the grid = 1 m in real life.

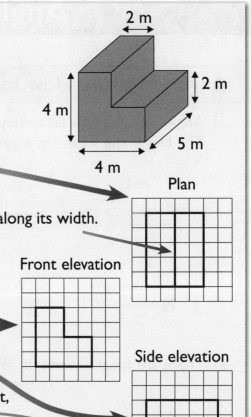

1) Start with the plan. If you look down on the object from above, it just looks like a rectangle. It is 4 m wide and 5 m long, so this is 4 squares wide and 5 squares long on the grid.

There's a change in the height of the shape 2 m along its width. Show this by drawing a line on your plan.

Plan

2) The front elevation will be an L-shape. It is 4 squares wide, 4 squares high on the left and 2 squares high on the right.

Front elevation

3) The side elevation will be another rectangle. It is 5 squares wide and 4 squares high.

The width of the shape changes 2 m up its height, so show this by drawing a line on the diagram.

Side elevation

Your drawings need to be accurate, so make sure you use a ruler and a sharp pencil.

Practice Question

1) Gwyn is planning to make a doll's house for his daughter. A sketch of the doll's house is shown on the right.

On the grids below, draw an accurate plan, front elevation and side elevation of the doll's house.

1 square on the grid = 10 cm in real life.

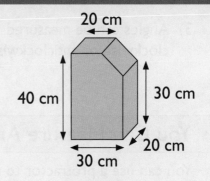

Plan

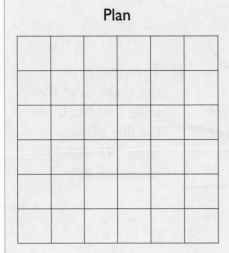

Front elevation

Side elevation

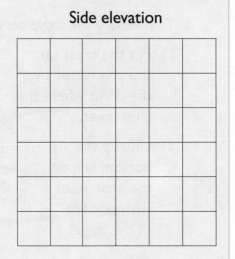

Angles and Bearings

Angles Measure How Far Something Has Turned

1) Angles tell you how far something has turned from a fixed point.
The bigger the angle, the bigger the turn. Angles are measured in degrees (°).

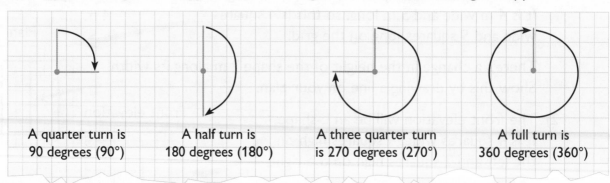

| A quarter turn is 90 degrees (90°) | A half turn is 180 degrees (180°) | A three quarter turn is 270 degrees (270°) | A full turn is 360 degrees (360°) |

2) There are special names for angles depending on their size.

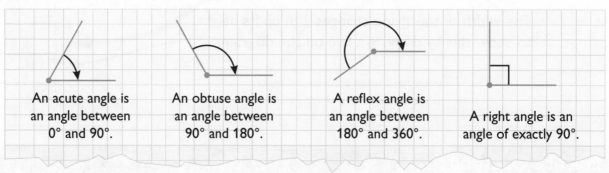

| An acute angle is an angle between 0° and 90°. | An obtuse angle is an angle between 90° and 180°. | A reflex angle is an angle between 180° and 360°. | A right angle is an angle of exactly 90°. |

3) Angles can be measured clockwise or anticlockwise.

Clockwise   Anticlockwise

You Can Measure Angles Between Lines

You can use a protractor to measure angles of up to 180°.

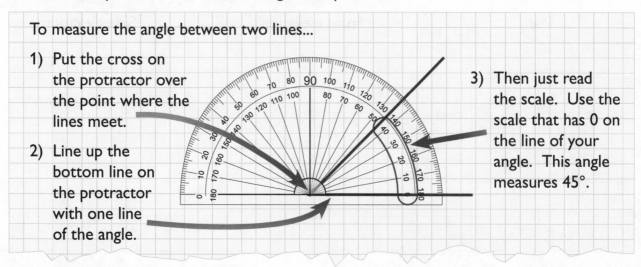

To measure the angle between two lines...

1) Put the cross on the protractor over the point where the lines meet.

2) Line up the bottom line on the protractor with one line of the angle.

3) Then just read the scale. Use the scale that has 0 on the line of your angle. This angle measures 45°.

Practice Questions

1) In the diagrams below, three arrows have been turned.
   Write the letter of the arrow next to the number of degrees it has turned.

   A 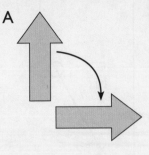   B   C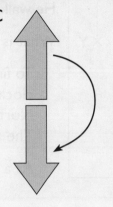

   90° = ...

   180° = ...

   270° = ...

2) Measure the angles between the lines below using a protractor. Write down the size of the angle (in degrees) and say whether it is acute, obtuse, reflex or a right angle.

   a)

   b)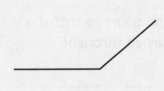

   c)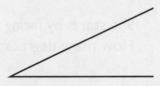

   ....

   ....

Bearings Measure the Angle of One Point From Another

1) A bearing is a direction given as an angle in degrees.

2) Bearings are always measured clockwise from a North line.

3) They are always given as three digits.
   For example, you write 025° instead of 25°
   and 061° instead of 61°.

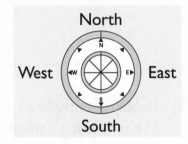

To find the bearing of B from A, draw a straight line connecting A and B.

Then measure the clockwise angle at A from North to the line.

The angle measures 110°, so the bearing is 110°.

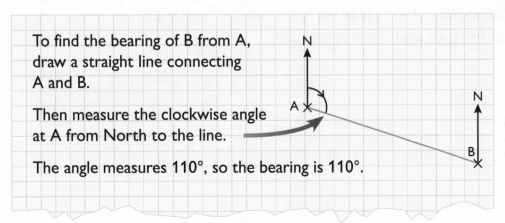

78

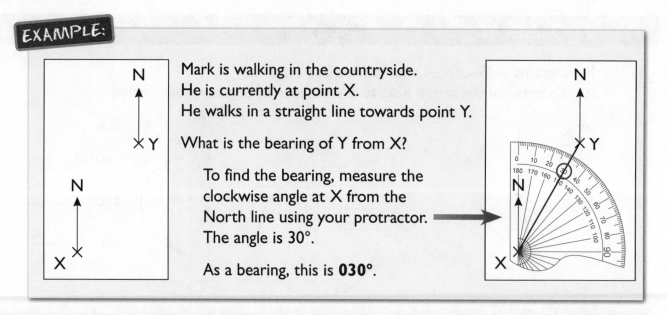

Practice Questions

1) Louise is testing out her new compass.

   She starts by facing North. She turns so that she is facing South.
   How many degrees has she turned through?

   ....

2) Find the bearing of V from U. You'll need to use your protractor.

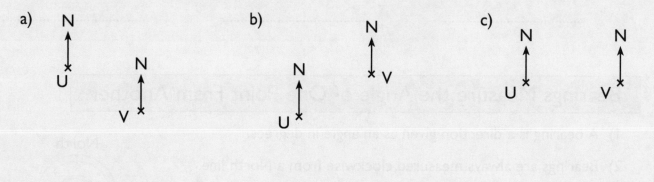

   ....

3) Bobbi is controlling a model car. It is currently at the point marked C.
   She moves it 12 cm on a bearing of 080° from C.

   Mark the new position of the car in the space below. Label it D.
   You'll need to use your protractor and a ruler.

Section Two — Measures, Shape and Space

Maps and Map Scales

You Need to Know How to Use a Map Scale

A map scale tells you how far a given distance on a map is in real life.

For example, a scale of 1 cm = 1 km means that 1 cm on the map equals 1 km in real life.

EXAMPLE 1:

Look at the map. What is the distance between Fleetley and Coneston in km?

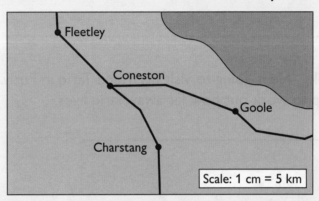

Scale: 1 cm = 5 km

1) Put your ruler against the bit you're finding the length of. Make sure the zero on the ruler is lined up with the starting place (in this case, Fleetley).

2) Mark off each whole cm and write the distance in km next to each one. In this case, 1 cm equals 5 km.

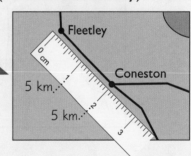

3) Add up all the km you just marked. So between Fleetley and Coneston:
5 km + 5 km = **10 km**.

EXAMPLE 2:

A map is drawn on a scale of 1 cm to 2 km.
If a road is 12 km long in real life, how long will it be in cm on the map?

Start by drawing the road as a straight line:

Mark off each cm and fill in how many km each one is:

Keep going until the km add up to the full distance (12 km in this case).

Then just count how many cm long your line is — in this case it's **6 cm**.

Practice Questions

1) A map is drawn with a scale of 1 cm to 4 km.

   a) If a road is 16 km in real life, how long will it be in cm on the map?

   ....

   ....

   b) If a road is 5 cm on the map, how long will it be in real life?

   ....

   ....

2) Simon lives in Oaks. He is going to visit his friend Tariq in Furly.
   He looks at a map to work out how far away Tariq lives.

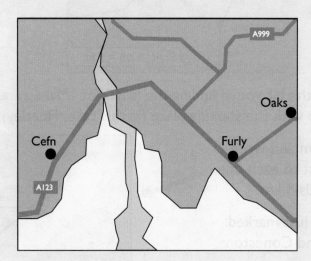

   Scale: 1 cm = 4 miles

   a) How many miles will he have to travel to get to Tariq's by road?

   ....

   ....

   b) Simon and Tariq decide to drive from Tariq's house to Cefn.
      They travel by road. How many miles is this journey?

   ....

   ....

   ....

   ....

You Might Have to Use Bearings and Maps Together

Questions might ask you to describe how far away something is and in which direction.

You'll have to use a map scale and measure the distance and the bearing.

EXAMPLE:

Bev is planning a trip from King's Lynn to Great Yarmouth.

Describe the position of Great Yarmouth from King's Lynn. Make sure to include:

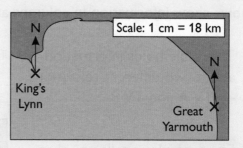

- the distance between them,
- the bearing of Great Yarmouth from King's Lynn.

1) Start with the distance. Draw a line between the towns and use your ruler to measure it. Count each whole centimetre and mark off the distance. Here, 1 cm represents 18 km.

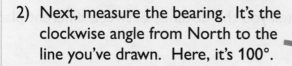

So the distance is 5 × 18 km = 90 km.

2) Next, measure the bearing. It's the clockwise angle from North to the line you've drawn. Here, it's 100°.

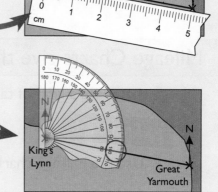

Great Yarmouth is 90 km from King's Lynn on a bearing of 100°.

Practice Question

1) Delia is sailing her boat on a lake. She is at the point marked D on this map. She needs to sail to the jetty at the point marked J.

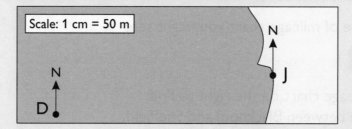

Scale: 1 cm = 50 m

Fully describe how to get directly from D to J. Make sure to give the distance and the bearing of J from D.

....

....

....

Tables

Tables are a Way of Showing Data

Tables show information in columns and rows.

This table holds information about two different televisions — TV A and TV B.

For example, it tells you that the price of TV A is £450 and the price of TV B is £550.

This table has row headings and column headings.

This is a column.

	TV A	TV B
Screen	37 inch	40 inch
HD	720p	1080p
Colour	Black	Silver
Price (£)	450	550

This is a row.

Mileage Charts give the Distances Between Places

Mileage charts are a kind of table. They tell you the distance between different places.

EXAMPLE 1:

Use the mileage chart to find the distance between Blackpool and Sheffield.

1) Move down the column underneath the 'Blackpool' heading and across the row next to the 'Sheffield' heading.

Blackpool			
52	Manchester		
56	34	Liverpool	
92	40	72	Sheffield

Distances are shown in miles.

2) Where the column and the row meet is the distance between the two places — **92 miles**.

Below is another type of mileage chart you might see.

EXAMPLE 2:

Use the mileage chart on the right to find the distance between Blackpool and Sheffield.

1) Find 'Blackpool' on one side of the chart and 'Sheffield' on the other.

2) Move across the row and down the column to the square where the row and the column meet — the distance is **92 miles**.

It doesn't matter if you've followed the path of the green arrows or the black ones, the answer will be the same.

	Blackpool	Manchester	Liverpool	Sheffield
Blackpool		52	56	92
Manchester	52		34	40
Liverpool	56	34		72
Sheffield	92	40	72	

Distances are shown in miles.

Practice Question

1) Use the mileage chart below to answer the following questions.
Distances are shown in miles.

London			
65	Bognor Regis		
292	354	Millom	
413	456	166	Edinburgh

a) What is the distance between Millom and London?

....

b) What is the distance between Bognor Regis and Edinburgh?

....

c) Jinden is driving from Edinburgh to Millom and then driving to London. What is the total mileage for his journey?

....

Tally Charts and Frequency Tables Show 'How Many'

You can use a tally chart or a frequency table to put data into different categories.

For example, this tally chart shows the colours of cars in a car park.

There are 2 blue cars.
There are 3 red cars.
There is 1 green car.
There are 6 silver cars.
There are 3 white cars.

Colour	Tally
Blue	II
Red	III
Green	I
Silver	IIII I
White	III

If another red car was seen in the car park you would add another line (tally mark) to the tally column next to red.

In a tally, every 5th mark crosses a group of 4 like this: IIII
So IIII I represents 6 (a group of 5 plus 1).

You can add another column to make the chart into a frequency table. You fill this in by adding up the tally marks for each colour.

Check the frequencies — the total should be the same as the number of tally marks (cars).

Colour	Tally	Frequency
Blue	II	2
Red	III	3
Green	I	1
Silver	IIII I	6
White	III	3
		Total: 15

Making a Table for a Set of Data

1) You might need to design your own table to collect or display data. There isn't just one right way to do this — it all depends on what data the table needs to show.

2) Make sure to include rows and columns for all the data, and enough space for everything that needs to be put into the table. (Check this again after you've drawn your table).

Laura is organising a dinner party for 8 guests. Some of her guests have special diets. The special diets are gluten-free and nut allergy.

Design a table to show each guest's diet and the total number of guests that have each type of diet. Some guests do not have a special diet — show this as well.

The table could look like this:

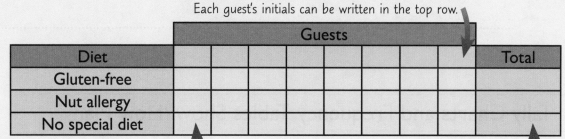

There's space for each of the 8 guests' diets to be shown. Each guest's initials can be written in the top row.

	Guests								Total
Diet									
Gluten-free									
Nut allergy									
No special diet									

The type of diet for each guest can be shown by putting a tick next to the diet type.

Include space for all the information — you're asked to include the total number of people with each diet so there needs to be somewhere to show that.

Practice Question

1) Terry is an electrician. He needs to place an order with his supplier.
Design a table that Terry could use to record the details of the order. The table should have space to record what items he is ordering, the number of each item he is ordering, the price of each item and the total cost of the order.

Charts and Graphs

Bar Charts Let You Compare Data Easily

1) A bar chart is a simple way of showing information.

2) On a bar chart you plot your data using two lines called axes
   (if you're talking about just one then it's called an axis).

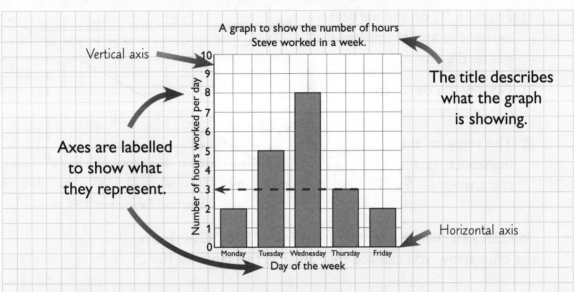

Vertical axis

A graph to show the number of hours
Steve worked in a week.

The title describes
what the graph
is showing.

Axes are labelled
to show what
they represent.

Number of hours worked per day

Monday Tuesday Wednesday Thursday Friday
Day of the week

Horizontal axis

1) The height of each bar shows how many hours were worked each day.

2) Just read across from the top of the bar to the number on the vertical axis.
   For example, Steve worked three hours on Thursday.

3) You can draw conclusions from the chart. For example, Steve worked
   the most hours on Wednesday as it's the day with the tallest bar.

Practice Question

1) The bar chart shows the number of different colours of
   shirt that were sold in a shop.

   a) How many blue shirts were sold?

   ....

   b) Which colour was the most popular?

   ....

   c) How many shirts were sold altogether?

   ....

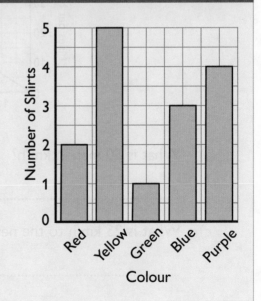

Number of Shirts

Red Yellow Green Blue Purple
Colour

Line Graphs Show the Relationship Between Two Things

Line graphs can be used to show how two lots of data are related.

EXAMPLE:

Here's a line graph for changing temperatures from degrees Celsius (°C) to degrees Fahrenheit (°F), or the other way round.

What is 35 °C in degrees Fahrenheit (°F)?

1) From 35 °C on the vertical axis, move across until you get to the line.

2) Go directly down to the horizontal axis.

3) The value on the horizontal axis is the answer — **95 °F.**

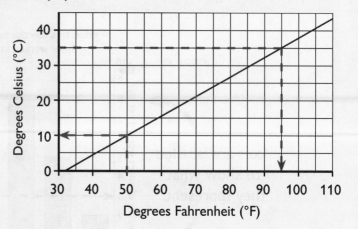

You can also change a temperature from °F to °C — from the blue arrow you can see that 50 °F is the same as 10 °C.

Practice Question

1) The graph below can be used to change between miles per hour (mph) and kilometres per hour (km/h).

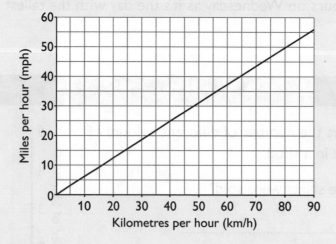

a) What is 50 mph in km/h?

b) What is 40 km/h in mph?

....

....

c) What is 25 km/h to the nearest mph?

....

Pie Charts

Pie Charts Show How Something is Split Up

1) Pie charts are circular and are divided into sections.

2) The size of each section depends on how much or how many of something it represents.

This pie chart shows the most popular equipment at a gym. The size of each section shows the proportion of people who prefer that piece of equipment.

This section is the biggest, so treadmills are the most popular piece of equipment.

It's half (50%) of the chart. This means that half of the people questioned prefer the treadmill.

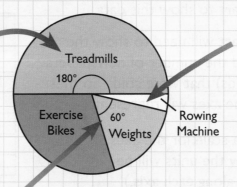

This is the smallest section on the chart.

This means that the rowing machine is less popular than other equipment at the gym.

This section has an angle of 60°. The angle all the way around the circle is 360°, and 60° is $\frac{1}{6}$ of 360°. So this means that $\frac{1}{6}$ of the people questioned prefer the weights.

See page 76 for how to use a protractor when you need to measure an angle.

Practice Question

1) The manager of a leisure centre is looking at the ages of members going to yoga and aerobics classes.

a) In which class are adults the most common age group?

....

b) What percentage of aerobics class members are youths?

....

c) The manager thinks that $\frac{1}{8}$ of the Yogo class is made up of youths. Is the manager correct?

....

....

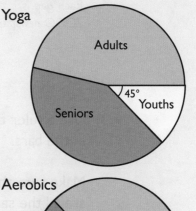

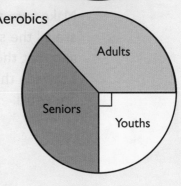

Drawing Charts, Graphs and Pie Charts

Drawing Bar Charts

You need to know how to draw a bar chart. The main steps are choosing what the axes will represent, choosing a scale for the axes and plotting (drawing) the data.

EXAMPLE:

The table below shows the monthly sales of a newspaper over 6 months. Draw a bar chart to show this data.

1) The bar chart will need to show the months and the number of newspapers (in thousands) that were sold. So these are what the axes will represent.

Month	Newspaper sales (thousands)
January	25
February	23
March	24
April	20
May	22
June	26

2) Work out a scale for the axes. (This is how the units will be spaced out along each axis.)

The biggest number of newspapers sold in a month is 26 000. So the axis needs to go from 0 to at least 26 000.

By giving the units as thousands you can just write, e.g. 26 on the scale.

Each gap is 2000 newspapers.

Add a title above your graph.

Make sure the axes are clearly labelled.

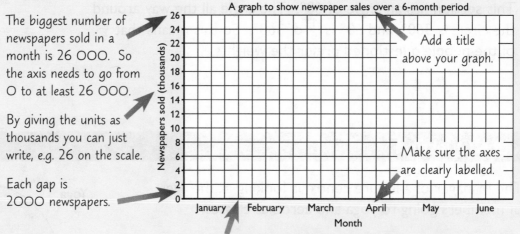

You need to pick a sensible scale for this axis. In this example, we've chosen to make each bar 2 squares wide, with a 2 square gap between them.

3) Use a ruler to draw in the bars.

Make sure the bars are all the same width and that the gaps between the bars are equal.

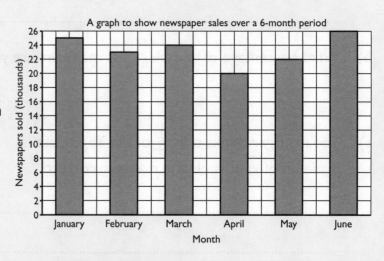

Drawing Line Graphs

The main steps for drawing line graphs are choosing what the axes will represent, choosing a scale for the axes and plotting the points.

EXAMPLE:

The growth of a tree over a period of 10 years is shown in the table.
Show this data on a line graph.

1) The line graph will need to show the year and the trunk diameter.

So the diameter of the trunk and the year are what the two axes will represent.

Year	0	2	4	6	8	10
Trunk diameter (inches)	3	5	7	8.5	10.5	12

2) Work out the scales for the axes.

Trunk diameter needs to go up to at least 12 inches — the highest diameter recorded.

Add labels to your axes. Make sure you include the units (inches).

If you use 1 square to represent 1 inch you need at least 12 squares for this axis.

Year needs to go from 0 to 10. Each square represents a year.

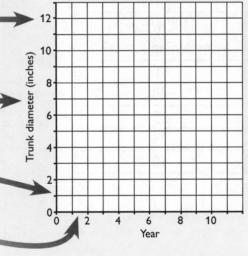

A graph to show the trunk diameter of a tree over 10 years.

3) Now plot the points.

For example, by year 4 the trunk diameter was 7 inches. Start at the year (4) and move up until you reach the diameter of the trunk (7) — draw a cross here.

Once you've plotted the points, join them with straight lines.

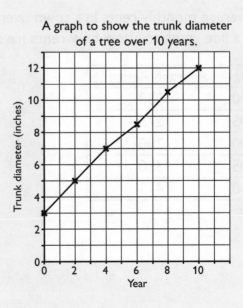

A graph to show the trunk diameter of a tree over 10 years.

Practice Questions

1) A company sells cars in different regions of the UK. The table shows the number of cars it has sold over the past 6 months. Draw a bar chart to display this information.

	Cars sold
North West	120
North East	140
Midlands	150

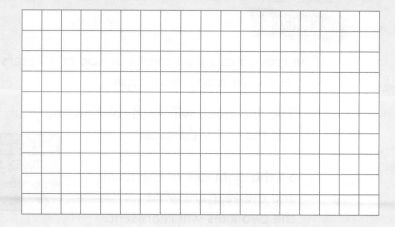

2) The amount of water left in the hot water tank of a house over the course of a day is shown in the table below. Draw a line graph to show how the amount of water changes.

Time	Hot water remaining (litres)
06:00	90
09:00	40
12:00	35
15:00	35
18:00	10

3) The average monthly rents in a town over 20 years are shown in the table below. Draw a line graph to show how rents have changed.

Year	Average Rent (£)
1990	250
1995	275
2000	300
2005	350
2010	475

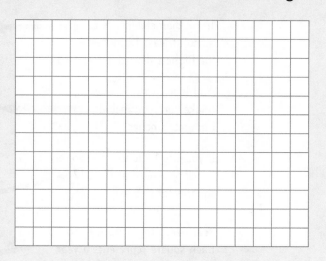

Drawing Pie Charts

1) You need to know how to draw pie charts.

2) Start by working out the angle of each section. Then accurately draw a circle using a pair of compasses and fill in the sections using a protractor.

EXAMPLE:

James works at a tourist information centre. The table shows the native languages of the tourists who visited the centre in the last month.

Language	Tourists
English	50
German	25
French	15

Draw a pie chart to display this data.

1) The pie chart will need three sections — one for each language spoken.

2) Divide 360° by the total number of tourists to find the angle that represents one tourist.

$\longrightarrow$ 50 + 25 + 15 = 90 tourists
360° ÷ 90 = 4°

3) Find the angle for each language by multiplying the angle of one tourist by the number of tourists who speak that language.

Language	English	German	French
Angle	50 × 4° = 200°	25 × 4° = 100°	15 × 4° = 60°

4) Draw a full circle and fill in each section accurately using a protractor.

5) Label each section with the thing that it represents.

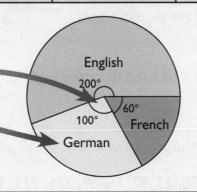

See page 76 for how to measure angles.

You don't need to shade the sections of your pie chart in the test.

Practice Question

1) The table below shows the type of 180 houses built in 2018.

Type	Houses	Angle
Bungalow	15	30°
Terrace	50	
Semi-detached	85	
Detached	30	

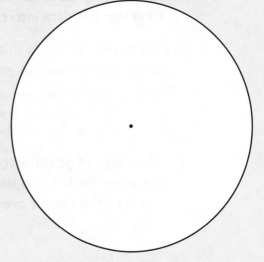

Complete the pie chart on the right to display this data.

Grouped Data

Grouping Data

1) You might need to group data into classes to make it more manageable.

2) Whatever data you have, make sure none of the classes overlap
and that they cover all of the possible values.

3) Tally charts can be used to collect grouped data.

See page 83 for more on tally charts and frequency tables.

EXAMPLE 1:

Amal is recording the ages (in whole years) of people who visit a museum.
Design a table that could be used to collect the data.

1) A tally chart is a good choice.

2) The data can be sensibly grouped
into five classes. There should be
no overlap in any of the classes.

3) Allow a class like '...or over' to
cover all of the possible ages.

Age (whole years)	Tally
0-19	
20-39	
40-59	
60-79	
80 or over	

There are lots
of other ways to
group this data.
For example,
you could use
0-24, 25-49, ...

4) A frequency table can be used to organise grouped data.

EXAMPLE 2:

Mona records the scores obtained by 30 people playing bowling.

127, 150, 23, 92, 66, 29, 271, 273, 189, 277, 164, 100, 133, 238, 47,
245, 58, 152, 203, 104, 117, 232, 110, 109, 24, 28, 158, 206, 83, 190

Create a table to organise this data.

1) A frequency table can be used
to group this data into classes.

2) The maximum score in a game is 300,
so the classes need to go from 0 to 300.

*If you didn't know the greatest possible score,
you could look at the highest actual score (277)
in the data and use this instead.*

3) The range of possible scores
has been divided into six classes.
None of the classes overlap.

The five scores of 23, 29, 47,
24 and 28 all lie in this class.

Score	Tally	Frequency
0-49	HHt	5
50-99	IIII	4
100-149	HHt II	7
150-199	HHt I	6
200-249	HHt	5
250-300	III	3
		Total: 30

The final class goes up to 300 (not 299)
so that all possible scores are covered.

Practice Question

1) Sally operates a private car park. She charges £1 per hour to park.
Below is a list of money raised for every day in the past month.

> £40, £21, £10, £83, £39, £28, £36, £34, £80, £57, £34, £15, £46, £78, £70,
> £45, £37, £57, £48, £13, £45, £23, £98, £41, £45, £79, £74, £36, £83, £28

Draw a grouped frequency table to organise this information.

Displaying Grouped Data

1) Grouped data can be displayed using bar charts.

> See page 88 for how
> to draw a bar chart.

2) There should be one bar to represent each class.

The table shows the age (in whole years) of athletes competing in a triathlon.

Age	Tally	Frequency
0-19	ＩＩＩＩ ＩＩ	7
20-29	ＩＩＩＩ ＩＩＩＩ ＩＩＩ	13
30-39	ＩＩＩＩ ＩＩＩＩ	9
40-49	ＩＩＩＩ	5
50 or over	ＩＩＩ	3
		Total: 37

A bar chart can be drawn to display the grouped data. It's just like drawing any other bar chart — put the classes on the horizontal axis and the frequency on the vertical axis.

The data has been grouped by age, so the bar chart is drawn with one bar for each age class.

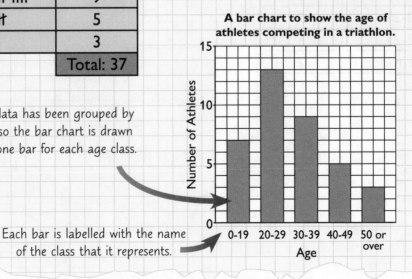

A bar chart to show the age of athletes competing in a triathlon.

Each bar is labelled with the name of the class that it represents.

Here's another way that data can be grouped.

EXAMPLE:

Mariam runs a bed and breakfast. The table shows the number of guests who stayed during each month of the year.

Group the data by season and display the grouped data in a bar chart.

Number of guests each month					
Jan	Feb	Mar	Apr	May	Jun
4	3	10	3	9	13
Jul	Aug	Sep	Oct	Nov	Dec
14	9	4	5	3	7

1) Create a new table for the grouped data.

Season	Number of guests
Spring (Mar-May)	10 + 3 + 9 = 22
Summer (Jun-Aug)	13 + 14 + 9 = 36
Autumn (Sep-Nov)	4 + 5 + 3 = 12
Winter (Dec-Feb)	7 + 4 + 3 = 14

2) Use the grouped data table to draw the bar chart.

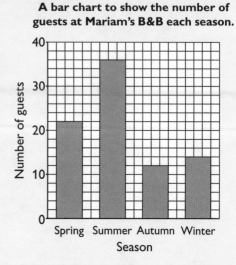

A bar chart to show the number of guests at Mariam's B&B each season.

Practice Question

1) Ash competes in an open water 800 m swim. The table below shows the finish times of the competitors to the nearest whole minute. Draw a bar chart to show these results.

Finish Time (minutes)	Frequency
17 or less	1
18-19	5
20-21	9
22-23	4
24 or more	2

Mean and Range

The Mean is a Type of Average

1) An average is a number that summarises a collection of data.

2) The average that you need to know how to work out is called the 'mean'.

> To work out the mean:
>
> 1) Add up all the numbers in the set of data.
>
> 2) Divide the total by how many numbers there are.

EXAMPLE 1:

The table shows the times taken by a group of women to run 100 m. What is the mean time taken?

1) First, add up the numbers:
   12.2 + 13.6 + 11.9 + 12.9 + 14.0 = 64.6

2) There are 5 numbers, so divide the total by 5:
   64.6 ÷ 5 = 12.92

3) The mean is **12.92 seconds**.

Runner	Time (s)
Jo	12.2
Rachel	13.6
Amélie	11.9
Samantha	12.9
Katrina	14.0

EXAMPLE 2:

Dave needs a new goal scorer for his football team.
He decides to move either Rafael, Jeremy or Paul up from the B team.

Dave decides to choose whoever scored the most goals on average in the last five matches. Who does he choose?

	Goals scored				
	Match 1	Match 2	Match 3	Match 4	Match 5
Rafael	Didn't play	2	3	0	1
Jeremy	4	Didn't play	0	4	0
Paul	Didn't play	0	Didn't play	2	1

Dave can work out the mean number of goals each player scored by dividing the total number of goals they scored by the number of games they played.

Rafael: 2 + 3 + 0 + 1 = 6   6 ÷ 4 = 1.5
Jeremy: 4 + 0 + 4 + 0 = 8   8 ÷ 4 = 2
Paul: 0 + 2 + 1 = 3   3 ÷ 3 = 1

Jeremy has the highest mean number of goals (2), so Dave will choose him.

Practice Questions

1) The total weight of 15 containers on a freight train is 330 tonnes.
   What is the mean weight of the containers?

....

2) Ellie asks her friends how far they have to travel to get to work. These are the results:

Distance travelled (miles)							
8	12	6	15	13	5	7	10

What is the mean distance travelled to work?

....

3) The table shows the price of a meal at six different restaurants.

Restaurant	A	B	C	D	E	F
Price	£23.10	£22.60	£23.70	£24.30	£26.20	£24.10

Calculate the mean price.

....

The Range is the Gap Between Biggest and Smallest

The range is the difference between the biggest value and the smallest value.

To work out the range:

1) Write down all the numbers in order from the smallest to the biggest.

2) Subtract the smallest number from the biggest number.

EXAMPLE 1:

The people waiting in a queue at a music festival are all asked their age.
The ages are: 18, 34, 18, 22, 20, 21, 26 and 24.
Work out the age range of the people waiting.

1) First, write the numbers in order of size:
   18, 18, 20, 21, 22, 24, 26, 34.

2) The biggest number is 34 and the smallest is 18.

Range = 34 – 18 = **16 years**.

If you know the range and one of either the smallest or largest values,
you can work out the other one of these values.

EXAMPLE 2:

Babies are weighed when they are born. In April, the weights of the babies
born at a hospital had a range of 1.9 kg. The lightest baby weighed 2.2 kg.
What was the weight of the heaviest baby?

The range is the difference between the heaviest and lightest weight.
So add the range to the lightest baby's weight:

The heaviest baby weighed 2.2 + 1.9 = **4.1 kg**.

You can check your answer: Range = heaviest − lightest = 4.1 − 2.2 = 1.9 ✓

Practice Questions

1) David has been recording the temperature in his greenhouse over the course of a week.
   Here are the temperatures he recorded at different times of the day:
   27 °C, 16 °C, 23 °C, 19 °C, 38 °C, 11 °C and 20 °C.

   a) What was the mean and the range of the temperatures that David recorded?

   ....

   ....

   b) David missed off 37 °C in his list. Would including this temperature change the range?

   ....

2) Asafa goes for a run every day after work for two weeks. He records the distance
   he covered each day in miles. They are: 2.2, 3.6, 2.9, 4.8, 4.6, 2.7, 5.2, 5.5, 4.3 and 3.7.

   a) What is the range of the distances that Asafa ran?

   ....

   ....

   b) Asafa recorded his longest run incorrectly. The correct distance is greater.
      After correcting the mistake, the range is 3.5 miles.
      What is the actual distance of Asafa's longest run?

   ....

   ....

Probability

Probability is all About Likelihood and Chance

1) Likelihood is how likely an event is to happen.

2) There are some key words you need to know:

- **Certain** — this is when something will definitely happen. For example, it's certain that you'll get a number from 1 to 6 when you roll a standard dice.

- **Likely** — this is when something isn't certain, but there's a high chance it will happen. For example, in a bag containing 99 blue counters and one red counter, you are likely to pick out a blue counter.

- **Even chance** — this is when something is as likely to happen as it is not to happen. For example, there's an even chance of getting heads when you toss a coin.

- **Unlikely** — this is when something isn't impossible, but it probably won't happen. For example, in a bag containing 99 blue counters and one red counter, you are unlikely to pick out a red counter.

- **Impossible** — this is when there's no chance at all of something happening. For example, it's impossible to roll a 7 on a standard six-sided dice.

3) An event being impossible isn't the same as one that is very, very unlikely. For example, it's very, very unlikely that it won't rain in the UK all winter, but it's not impossible.

Practice Question

1) Describe the following events as certain, likely, even chance, unlikely or impossible.

a) Your mother is younger than you.

....

b) New Year's Day will be on the 1st of January next year.

....

c) The weather will be hot and sunny on 25th December in the UK.

....

Probability can be Shown on a Scale

Probabilities can be shown on a scale from 0 to 1 like the one below.

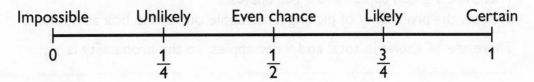

Impossible	Unlikely	Even chance	Likely	Certain
0	$\frac{1}{4}$	$\frac{1}{2}$	$\frac{3}{4}$	1

EXAMPLE:

What is the probability of getting a head when tossing a coin?

There is an even chance of getting a head or a tail, so the probability is $\frac{1}{2}$.
This can be shown on a probability scale:

Probability of
getting a head

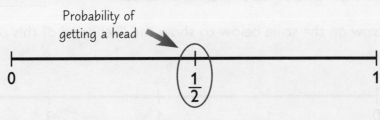

0        $\frac{1}{2}$ 1

You May Need to Calculate a Probability

1) You can calculate the probability that something will happen.
   Your answer should be a fraction between 0 and 1.

2) To calculate a probability, divide the number of ways that the thing can happen
   by the total number of possible outcomes.

$$\text{Probability} = \frac{\text{Number of ways for something to happen}}{\text{Total number of possible outcomes}}$$

An 'outcome' is just one thing that could happen.

EXAMPLE 1:

Sam is a hockey club coach. His team has 16 players.
He divides the team by asking each player to draw a ticket at random from a hat.
The hat contains 2 blue, 3 green, 5 red and 6 yellow tickets.
What is the probability of the first player to pick getting a red ticket?

There are 5 red tickets and

16 tickets in total, so the probability is $\frac{5}{16}$.

There are 5 ways of picking a red ticket, since there are 5 red tickets in the hat.

There are 16 possible outcomes, since there are 16 tickets in total in the hat.

EXAMPLE 2:

Mohini has brought a box of apples into work to share with her colleagues.
There are 5 green apples and 9 red apples.
What is the probability of picking a red apple out of the box at random?

There are 14 apples in total and 9 red apples, so the probability is $\frac{9}{14}$.

Practice Questions

1) Gavin has 4 pairs of socks — 3 red pairs and 1 black pair.
   This morning, he picked a pair of socks to wear without looking.

   a) Describe the likelihood of Gavin picking a red pair. ....

   b) Draw an arrow on the scale below to show the likelihood of this outcome.

   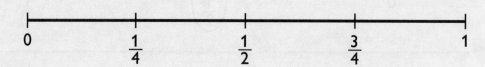

   0        $\frac{1}{4}$ $\frac{1}{2}$ $\frac{3}{4}$ 1

2) Sophia has a pack of buttons containing 4 blue, 6 pink, 8 white and 5 red buttons.
   She picks one out without looking. Work out the probability of Sophia picking a red button.

   ....

3) Simon has baked 90 pies. 30 are cheese and onion, 30 are meat and potato and
   the others are steak and blue cheese. They are all mixed up on a tray. If he picks
   one at random, what is the probability it will be a steak and blue cheese pie?

   ....

4) A travel company has a fleet of 16 coaches. 12 are painted black and 4 are painted white.
   At the start of the day, all the coaches are available and Geoff picks up a set of coach keys.

   a) What is the probability he has the keys to a white coach?

   ....

   b) What is the probability he has the keys to a black coach?

   ....

5) Rachel has made ten necklaces to sell at a craft fair. 8 of them have a quartz set into them
   and the other two have moonstones instead. Whilst unpacking them, what is the chance
   that Rachel picks a quartz necklace out first?

   ....

Candidate Surname	Candidate Forename(s)

Test Date	Candidate Signature

Functional Skills

Mathematics   Level 1

Section A — Non-calculator
Time allowed: 25 minutes

You **may not** use a calculator.

There are **14 marks** available for this section.

Section B — Calculator
Time allowed: 1 hour 30 minutes

You **may** use a calculator.

There are **42 marks** available for this section.

You must have:
Pen, pencil, eraser, ruler, protractor, compasses, calculator (Section B only).

Instructions to candidates

- Use **black ink** to write your answers.
- Write your name and the date in the spaces provided above.
- There are **2 sections** in this paper.
  Answer **all questions** in each section in the spaces provided.
- In calculations, show clearly how you worked out your answers.
- Check your working and answers.

Information for candidates

- Diagrams are **not** accurately drawn, unless otherwise stated.
- The marks available are given in brackets at the end of each question.
- Marks will be awarded for **showing your check** when you see this symbol ☑.

Section A — Non-calculator

Answer all questions in this section.

Write your answers in the spaces provided.

1 Lola is making fruit cakes to sell at a bake sale.
She uses two recipes that require different amounts of flour.

Recipe A	Recipe B
1.98 kg of flour	0.495 kg of flour

She makes three cakes using Recipe A and two cakes using Recipe B.

(a) Estimate how much flour she used in total.

(3)

kg

The money raised at the bake sale is donated to charity.
£3 out of every £5 goes to a children's hospital.

(b) What is this as a percentage?

(2)

%

(Total for Question 1 is 5 marks)

2 Josh is constructing a stage at a music concert.
The stage is in the shape of a cuboid. It is 2 m high.

2 m

The stage is 4 m long and 12 m wide.

(a) What is the volume of the stage?
Remember to give units with your answer.

(2)

☑ **(b)** Use reverse calculations to check your answer.

(1)

(Total for Question 2 is 3 marks)

3

(a) Calculate 198 ÷ 1000.

(1)

(b) Calculate 6 + 10 ÷ 2.

(1)

(c) What fraction is $2\frac{3}{5}$ equal to?

(1)

5

(Total for Question 3 is 3 marks)

4   The table below shows the money spent in a garden centre each day of last week.

Day	Amount spent (in £)
Mon	1180
Tue	1670
Wed	3300
Thur	912
Fri	2310
Sat	4290

(a) Work out the range of the amounts spent.

(2)

£

$\frac{1}{3}$ of the money spent on Thursday was from plant sales.

(b) How much money is this?

(1)

£

(Total for Question 4 is 3 marks)

End of Section A

Section B — Calculator

Answer all questions in this section.

Write your answers in the spaces provided.

1   Helen uses the map below to plan a bike ride.
    It is drawn accurately.

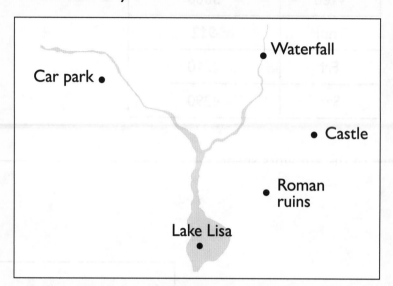

Scale: 1 cm on the map is 2 km in real life.

She cycles in a straight line from the car park to the waterfall and then to the castle.

(a) How far does she cycle in total?

(3)

km

The diagram below shows a close-up of part of the map.
It is drawn accurately.

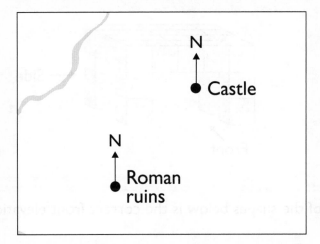

Salem is at the Roman ruins. He wants to meet Helen at the castle.

(b) What is the bearing of the castle from the Roman ruins?

(1)

°

(Total for Question 1 is 4 marks)

2 Deangelo has bought a table, shown below.
It is made from identical cubes.

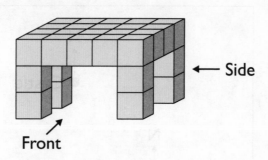

Side ←

↗ Front

(a) Which of the shapes below is the correct front elevation?

(1)

Shape A   Shape B   Shape C   Shape D

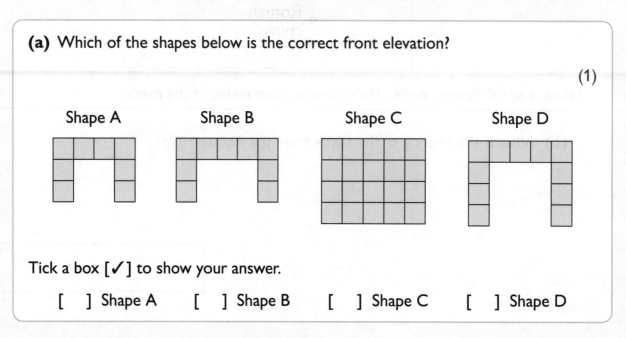

Tick a box [✓] to show your answer.

[   ] Shape A   [   ] Shape B   [   ] Shape C   [   ] Shape D

The side elevation of the table is drawn below.

0.2 m

0.2 m

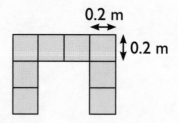

(b) What is the height of the table?

(2)

m

(Total for Question 2 is 3 marks)

3   The table below shows the feeding times of some animals in a zoo.
    Each feeding session lasts for 20 minutes.

Animal	Feeding times
Giraffes	12:00
Lions	11:30 and 13:30
Penguins	12:30
Vultures	10:30 and 14:00
Lemurs	15:00

Marlon watches the lions feed in the morning.
He then spends three quarters of an hour walking around before stopping for lunch.

At what time did Marlon stop for lunch?
Give your answer in the 24-hour clock.

(2)

(Total for Question 3 is 2 marks)

4   Alicia buys a new car that has driven 0 miles.
    She records the distance she drives each year for the next four years.
    This table shows her results.

	Year 1	Year 2	Year 3	Year 4
Distance (miles)	9805	7106	8504	9593

(a) What is the mean distance driven in a year?

(3)

miles

Alicia bought the car for twelve thousand, five hundred pounds.
Since then, its value has decreased by 45%.

(b) How much is the car worth now?

(3)

£

(Total for Question 4 is 6 marks)

5 The receipt below shows what Emma bought at the shop one morning.

cereal	2 × £2.90 = £5.80
sandwich	£2.99
crisps	45p
drink	60p

There are two offers that Emma could have used to save money.

Meal Deal
Any sandwich, crisps and drink for £3

Voucher
20% off all cereal

Emma used the offer that saved her the most money.
She could not use both offers.

Which offer did Emma use?
You **must** show your working.

(4)

Tick a box [✓] to show your answer.

[   ] Meal Deal   [   ] Voucher

(Total for Question 5 is 4 marks)

6 Tony's bookshop receives a delivery of new books once a month.
He records this information about the books:

New fiction books for July		
Category	Number of books	Angle in pie chart
Crime	16	80°
Romance	24	120°
Historical	20	
Sci-Fi	12	
	Total: 72	

Tony has started to construct a pie chart to display this information.

Complete Tony's pie chart.

(3)

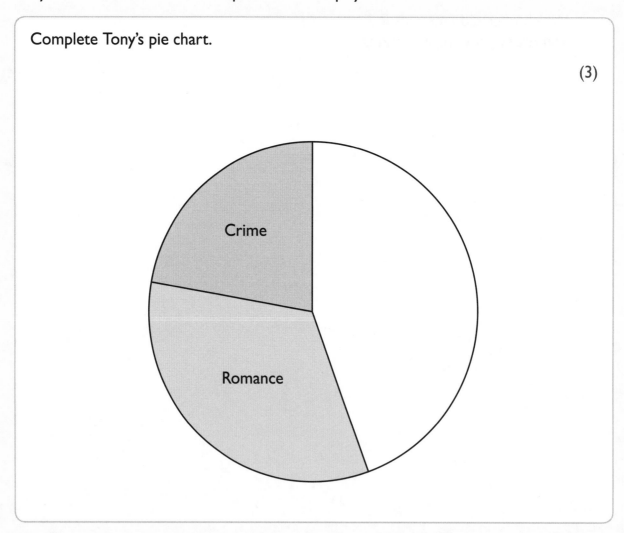

(Total for Question 6 is 3 marks)

7 Inga is organising a surprise party for her sister Sofia's birthday.
Sofia usually leaves work at 5:30 pm.
The arrow on the scale shows the probability of Sofia arriving home before 6 pm.

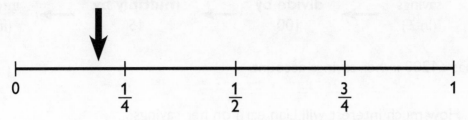

(a) Describe the probability of Sofia arriving home before 6 pm.

(1)

Tick a box [✓] to show your answer.

[   ] Certain

[   ] Likely

[   ] Even chance

[   ] Unlikely

[   ] Impossible

Inga makes 25 party invitations out of coloured paper.
10 of the invitations are red, 8 are blue and the rest are green.

She picks one up at random and posts it.

(b) What is the probability that Inga posted a green invitation?

(2)

(Total for Question 7 is 3 marks)

8 Lian has a savings account with her bank.
The bank uses the rule below to calculate how much interest is earned.

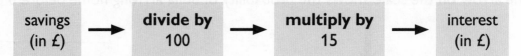

| savings (in £) | → | **divide by** 100 | → | **multiply by** 15 | → | interest (in £) |

Lian has £1200 in her savings account.

(a) How much interest will Lian earn on her savings?

(2)

£ [   ]

☑ **(b)** Use reverse calculations to check your answer.

(1)

(Total for Question 8 is 3 marks)

9 Aftab wants a patio in his garden.
He is going to use concrete slabs to make the patio.

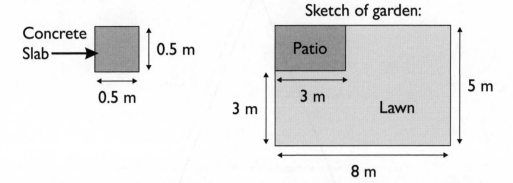

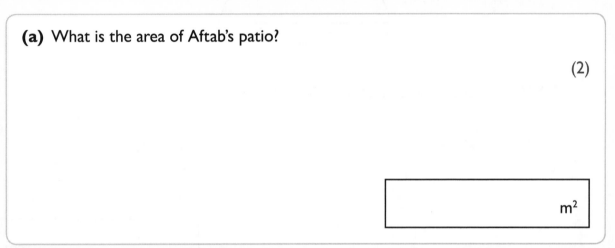

(a) What is the area of Aftab's patio?

(2)

m²

Aftab pays £3.10 per concrete slab.

(b) How much will it cost Aftab to buy enough concrete slabs to make the patio?

(3)

£

(Total for Question 9 is 5 marks)

10 The shape below is an isosceles triangle.

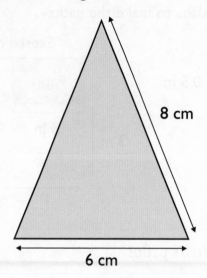

8 cm

6 cm

(a) What is the perimeter of the triangle?

(2)

cm

(b) How many lines of symmetry does the triangle have?

(1)

(Total for Question 10 is 3 marks)

11 Davina works at a call centre.
The time taken to answer each call is recorded in whole seconds.

This morning, Davina answered 6 calls within 15-29 seconds
and two thirds as many within 30-44 seconds.

An incomplete bar chart has been drawn below to show how long
it took Davina to answer each call this morning.

(a) Use the information above to complete the bar chart.

(2)

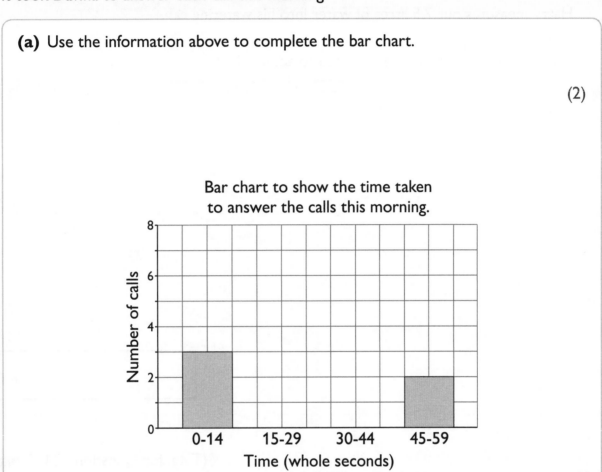

Bar chart to show the time taken
to answer the calls this morning.

Davina is supposed to answer each call in fewer than 30 seconds.

(b) How many of this morning's calls did Davina answer in fewer than 30 seconds?

(1)

calls

(Total for Question 11 is 3 marks)

12 Harry has bought some liquid lawn feed to put on his lawn.
He reads the instructions to find out how to dilute the lawn feed.

To Use
Lawn feed must be diluted with water.
For each 5 ml of lawn feed, add 75 ml of water.

Harry measures out 7.5 litres of water into his watering can.

How much lawn feed does he need to add to this?

(3)

| | ml |

(Total for Question 12 is 3 marks)

End of Section B

Answers — Practice Questions

Section One — Number

Page 4
Q1 Five hundred and three thousand, four hundred and twenty-five
Q2 a) 165   b) 671 902
Q3 Amelia, Yuri, Rosa, Neil, Marie, Marco

Page 6
Q1 −8
Q2 In the morning
Q3 a) 23 °C   b) 27 °C

Page 7
Q1 −1 °C
Q2 −5 °C
Q3 12 °C

Page 9
Q1 27 470 steps
Q2 750 g
Q3 £15 780

Page 12
Q1 240
Q2 23
Q3 a) 19 boxes with 4 left over
    b) 1114

Page 13
Q1 363 ÷ 11 = 33 or 363 ÷ 33 = 11
Q2 1602 − 803 = 799 (not 899)
    or 1602 − 899 = 703 (not 803)
    or 803 + 899 = 1702 (not 1602)
    So Mark is incorrect.
Q3 6 × 24 = 144 or 144 ÷ 6 = 24
    So Allison is correct.

Page 14
Q1 a) 230   b) 46.5
    c) 789   d) 9700
Q2 a) 65 700   b) 460
    c) 9780   d) 2 980 000
Q3 8850 m

Page 15
Q1 a) 1.8   b) 0.48
    c) 465.8   d) 9.72
Q2 a) 4.159   b) 0.024
    c) 3.214   d) 2.678
Q3 44.3 miles

Page 16
Q1 a) 36   b) 9
    c) 144
Q2 529

Page 18
Q1 14
Q2 a) 6   b) 12

Q3 3
Q4 1
Q5 a) 21   b) 5

Page 19
Q1 $\frac{3}{5}$
Q2 $\frac{3}{8}$

Page 20
Q1 a) $\frac{9}{5}$ b) $\frac{13}{6}$
    c) $\frac{26}{7}$ d) $\frac{27}{10}$
Q2 a) $1\frac{2}{3}$ b) $3\frac{3}{4}$
    c) $3\frac{3}{5}$ d) $2\frac{5}{8}$

Page 22
Q1 15
Q2 4
Q3 $\frac{10}{24}$
Q4 a) $\frac{4}{8}, \frac{6}{10}, \frac{5}{8}$ b) $\frac{19}{20}, 1\frac{1}{5}, 1\frac{1}{4}$
Q5 Rachel

Page 23
Q1 6
Q2 20
Q3 a) 14 hours   b) 10.5 hours

Page 25
Q1 a) 1.556   b) 3.145
    c) 46.58
Q2 a) 0.658   b) 15.604
    c) 6.431
Q3 0.6 kg, 0.603 kg, 0.63 kg, 6.006 kg

Page 27
Q1 12.14
Q2 2.85 cm
Q3 £12.99 + £4.95 + £1.62 = £19.56.
    This is less than £20, so Matt will not get a free watering can.

Page 29
Q1 4.272
Q2 15.6
Q3 a) 5
    b) No (he only has enough paint for 14.25 m²)

Page 31
Q1 a) 3.48   b) 20.46
Q2 a) 0.9   b) 6.6
Q3 a) 45.80   b) 3.0
Q4 a) 51   b) 6
Q5 35 °C

Page 32
Q1 a) 30 + 100 + 1000 = 1130
    b) 40 × 2 = 80
    c) (200 − 60) ÷ 7 = 20
Q2 For example:
    £30 + £30 + £10 + £20 + £100 = £190

Page 34
Q1 $\frac{35}{100}$
Q2 7.5
Q3 a) 16   b) 13

Page 35
Q1 36
Q2 13
Q3 16
Q4 180
Q5 88

Page 36
Q1 a) 531 miles   b) 4071 miles
Q2 3.3 m
Q3 57 minutes

Page 38
Q1 0.05
Q2 $\frac{1}{4}$
Q3 1.25 kg
Q4 a) 0.6   b) 60%
Q5 a) 80%   b) Cumbria

Page 39
Q1 0.25
Q2 Leanne pays £400. Tamal pays £360. So Tamal pays less for a new TV.

Page 41
Q1 a) 1:4   b) 100 ml
Q2 18
Q3 £2000 and £1000

Page 43
Q1 48 kg
Q2 750 ml
Q3 a) 4 eggs   b) 360 g
Q4 £1500
Q5 6

Page 46
Q1 £57
Q2 a) 120 minutes (or 2 hours)
    b) 90 minutes (or 1 hour 30 minutes)
Q3 £312
Q4 Less, 32 m² is smaller than the maximum area of 32.5 m².

Section Two — Measures, Shape and Space

Page 47
Q1 a) 127p   b) £2.19
Q2 £25.59

Page 49
Q1 £3.50
Q2 £2560
Q3 £39
Q4 £412.50
Q5 £6500
Q6 Shop B

Page 51
Q1 a) 200 cm   b) 5 cm
     c) 150 cm
Q2 5 km
Q3 170 cm
Q4 235 km or 235 000 m
Q5 Yes, the total width of the furniture is 125 + 90 = 215 cm, while the width of the wall is 2.3 × 100 = 230 cm.

Page 53
Q1 1.066 kg
Q2 4
Q3 Yes, Damon and his equipment weigh a total of 56.4 kg.
Q4 6
Q5 200

Page 55
Q1 0.5 L
Q2 5
Q3 285 ml
Q4 No, the ingredients add up to 450 ml which is more than 400 ml.

Page 57
Q1 120 minutes
Q2 30 years
Q3 4 minutes
Q4 a) 9:00 am   b) 4:45 pm
Q5 a) 17:15   b) 07:05
Q6 No (10:47 pm is 22:47).

Page 59
Q1 2 hours 15 mins (or 135 mins)
Q2 2 hours 52 minutes (or 172 mins)
Q3 7:50 pm (or 19:50)
Q4 1 hour 55 minutes (or 115 mins)
Q5 20:15 (or 8:15 pm)
Q6 Yes (if it takes a quarter of an hour to find a parking space he should get to the theatre at 19:15).

Page 60
Q1 a) 18 cm   b) 26 cm

Page 62
Q1 a) 24 m   b) 44 cm
Q2 a) 24 cm (the unknown side is 4 cm long)

     b) 28 cm (the unknown sides are 4 cm and 3 cm long)
     c) 24 cm (the unknown sides are 4 cm and 2 cm long)
Q3 13.25 m

Page 64
Q1 a) 6 cm²   b) 42 cm²
     c) 30 cm²   d) 32 cm²

Page 66
Q1 £693
Q2 96 tiles

Page 68
Q1 a) 12 cm³   b) 10 cm³
Q2 a) 24 cm³   b) 50 cm³
Q3 0.16 m³

Page 71
Q1 a)

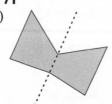

b)

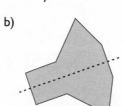

c)

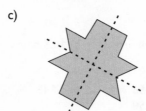

Q2 a)

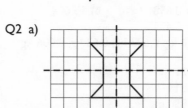

b)

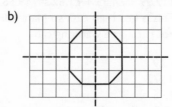

Q3 a) 9.42 cm   b) 3 cm

Q4 a)

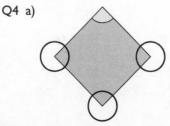

b)

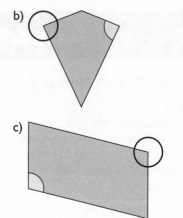

c)

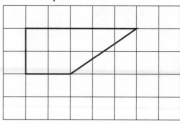

Q5 For example:

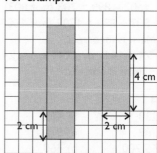

Page 74
Q1 For example:

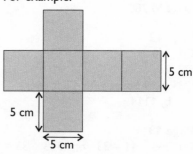

5 cm
5 cm
5 cm

Q2 For example:

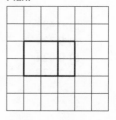

4 cm
2 cm   2 cm

Page 75
Q1 Plan:

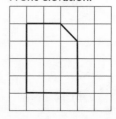

Front elevation:

Side elevation:

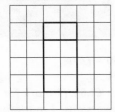

Page 77
Q1 90° = A
180° = C
270° = B
Q2 a) 80°, acute   b) 140°, obtuse
c) 25°, acute

Page 78
Q1 180°
Q2 a) 120°   b) 060°
c) 090°
Q3

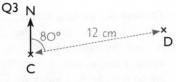

Page 80
Q1 a) 4 cm   b) 20 km
Q2 a) 8 miles   b) 26 miles

Page 81
Q1 J is 300 m from D on a
bearing of 080°.

Section Three —
Handling Data

Page 83
Q1 a) 292 miles   b) 456 miles
c) 458 miles

Page 84
Q1 For example:

Item	Quantity	Price	Total cost of items
		Total cost of order	

There are other ways of drawing this
table. Just make sure you've left space
for all of the details you were asked for.

Page 85
Q1 a) 3   b) Yellow
c) 15

Page 86
Q1 a) 80 km/h   b) 25 mph
c) 15 mph or 16 mph

Page 87
Q1 a) Yoga   b) 25%
c) Yes (45° is $\frac{1}{8}$ of 360°)

Page 90
For these questions, your graphs may
not look exactly like the ones drawn
here. For example, you might have
used a different scale for your axes.
Q1

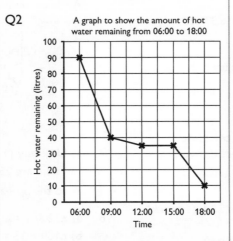

Q2

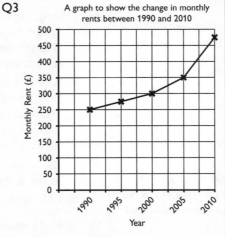

Q3

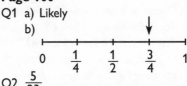

Page 91
Q1

Terrace 100°
Bungalow
170°
30°
60°
Semi-detached
Detached

Page 93
Q1 For example:

Money (£)	Tally	Frequency
0-19	III	3
20-39	JHT JHT	10
40-59	JHT IIII	9
60-79	IIII	4
80-99	IIII	4
		Total: 30

Page 94
Q1

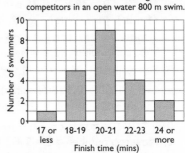

Page 96
Q1 22 tonnes
Q2 9.5 miles
Q3 £24

Page 97
Q1 a) Mean = 22 °C Range = 27 °C
b) No (it's between 11 and 38 °C).
Q2 a) 3.3 miles   b) 5.7 miles

Page 98
Q1 a) Impossible   b) Certain
c) Unlikely

Page 100
Q1 a) Likely
b)

0    $\frac{1}{4}$ $\frac{1}{2}$ $\frac{3}{4}$ 1

Q2 $\frac{5}{23}$

Q3 $\frac{30}{90}$ (or any equivalent fraction,
for example $\frac{3}{9}$ or $\frac{1}{3}$)

Q4 a) $\frac{4}{16}$ (or any equivalent fraction,
for example $\frac{2}{8}$ or $\frac{1}{4}$)

b) $\frac{12}{16}$ (or any equivalent fraction,
for example $\frac{6}{8}$ or $\frac{3}{4}$)

Q5 $\frac{8}{10}$ (or any equivalent fraction,
for example $\frac{4}{5}$)

Answers

Answers — Practice Paper

Section A — Non-calculator (Page 102)

1 a) 1.98 = 2.0 to 1 d.p. and 0.495 = 0.5 to 1 d.p. *(1 mark)*
   2 × 3 = 6 and 0.5 × 2 = 1 *(1 mark)*
   So she uses about 6 + 1 = **7 kg** of flour. *(1 mark)*

   b) $\frac{3}{5}$ = 3 ÷ 5 = 0.6 *(1 mark)*, 0.6 × 100 = 60
   So £3 out of every £5 is **60%**. *(1 mark)*

2 a) 12 × 4 × 2 = 48 × 2 = **96 m³**
   (1 mark for multiplying 12, 4 and 2,
   1 mark for the correct answer with units)
   You could multiply 12, 4 and 2 in any order.

   b) For example: 96 ÷ 4 = 24 and 24 ÷ 2 = 12 *(1 mark)*

3 a) 198 ÷ 1000 = **0.198** *(1 mark)*
   You move the decimal point 3 places to the left.

   b) 6 + 10 ÷ 2 = 6 + 5 = **11** *(1 mark)*

   c) New top number = 2 × 5 + 3 = 10 + 3 = 13
   So $2\frac{3}{5} = \frac{13}{5}$. *(1 mark)*

4 a) Range = biggest value – smallest value
   4290
   – 912
   3378   So the range is **£3378**.
   (1 mark for a correct method to calculate
   4290 – 912, 1 mark for the correct answer)

   b) $\frac{1}{3}$ of 912 = 912 ÷ 3:   304
   So the money spent on plants was **£304**. *(1 mark)*

Section B — Calculator (Page 106)

1 a) Distance from car park to waterfall = 4.5 cm
   Distance from waterfall to castle = 2.5 cm *(1 mark)*
   Total distance = 4.5 + 2.5 = 7 cm *(1 mark)*
   7 cm × 2 = **14 km** *(1 mark)*
   Use a ruler to accurately measure the distances between
   the points in the diagram. Answers between 13.6 km
   and 14.4 km are acceptable.

   b) **040°** *(1 mark)* Accept answers between O38° and O42°.

2 a) **Shape B** *(1 mark)*
   b) The table is 3 cubes tall and each square has a height
   of 0.2 m. So the table has a height of 3 × 0.2 = **0.6 m**.
   (1 mark for 3 × 0.2, 1 mark for the correct answer)

3 For example: 11:30 + 20 minutes = 11:50
                 11:50 + 45 minutes = 12:35
   So Marlon stopped for lunch at **12:35**.
   (1 mark for using a correct method, 1 mark for the
   correct answer)

4 a) Total distance = 9805 + 7106 + 8504 + 9593
                     = 35 008 miles *(1 mark)*
   Mean = 35 008 ÷ 4 *(1 mark)* = **8752 miles** *(1 mark)*

   b) 45% of £12 500 = $\frac{45}{100}$ × 12 500
                     = 45 ÷ 100 × 12 500 = £5625
   New value = 12 500 – 5625 = **£6875**
   (1 mark for using 12 500, 1 mark for a correct
   method of dealing with percentages, 1 mark for
   the correct answer)

5 Meal deal: £2.99 + £0.45 + £0.60 = £4.04,
   £4.04 – £3.00 = £1.04
   So the meal deal saves £1.04. *(1 mark for*
   correct currency conversions, 1 mark for
   the correct answer)

   Voucher: 20% of £5.80 = $\frac{20}{100}$ × 5.80
                          = 20 ÷ 100 × 5.80 = £1.16
   So the voucher saves £1.16. *(1 mark)*
   £1.16 is greater than £1.04, so Emma used
   the **voucher**. *(1 mark)*
   You must state your final answer and write down your
   reasoning to get full marks.

6 Angle for one book = 360° ÷ 72 = 5°
   Angle for historical books = 20 × 5° = **100°**
   Angle for sci-fi books = 12 × 5° = **60°**

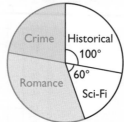

   (1 mark for using a correct
   method to calculate angles,
   1 mark for one or both
   correct angles shown in table
   or on pie chart, 1 mark for
   the completed pie chart
   with both correct angles)

7 a) **Unlikely** *(1 mark)*
   b) There are 25 – 10 – 8 = 7 green invitations *(1 mark)*,
   so the probability is $\frac{7}{25}$. *(1 mark)*

8 a) £1200 ÷ 100 = £12 *(1 mark)*, £12 × 15 = **£180** *(1 mark)*
   b) £180 ÷ 15 = £12, £12 × 100 = **£1200** *(1 mark)*

9 a) Length of the missing side = 5 – 3 = 2 m
   Area = 2 × 3 = **6 m²**
   (1 mark for correct area formula, 1 mark for the
   correct answer)
   b) Area of a slab = 0.5 × 0.5 = 0.25 m² *(1 mark)*
   So Aftab needs 6 ÷ 0.25 = 24 slabs. *(1 mark)*
   The total cost is 24 × £3.10 = **£74.40**. *(1 mark)*
   You have to use the correct money format to get the
   last mark. So writing '£74.4' wouldn't be right.

10 a) The length of the missing side is 8 cm. *(1 mark)*
   So the perimeter is 6 + 8 + 8 = **22 cm**. *(1 mark)*
   b) **1** *(1 mark)* It's vertically down the middle.

11 a) 30-44 seconds: $\frac{2}{3}$ of 6 = 6 ÷ 3 × 2 = 2 × 2 = 4 calls

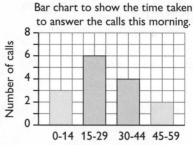

   (1 mark for
   each correctly
   drawn bar)

   b) 3 + 6 = **9 calls** *(1 mark)*

12 7.5 × 1000 = 7500 ml *(1 mark)*
   7500 ml ÷ 75 ml = 100 *(1 mark)*
   100 × 5 ml = **500 ml** *(1 mark)*

Glossary

12-hour Clock

The 12-hour clock goes from 12:00 am (midnight) to 11:59 am (one minute before noon), and then from 12:00 pm (noon) to 11:59 pm (one minute before midnight).

24-hour Clock

The 24-hour clock goes from 00:00 (midnight) to 23:59 (one minute before the next midnight).

2D Object

An object with 2 dimensions, i.e. a flat object.

3D Object

An object with 3 dimensions, i.e. a solid object.

 A

Angle

A measurement of how far something has turned from a fixed point.

Anticlockwise

Movement in the opposite direction to the hands of a clock.

Area

How much surface a shape covers.

Average

A number that summarises a lot of data.

Axis

A line along the bottom and up the left-hand side of most graphs and charts. The plural is 'axes'.

 B

Bar Chart

A chart which shows information using bars of different heights.

Bearing

A direction given as an angle. It is measured clockwise from North and written in degrees as three digits. For example, 070°.

BIDMAS

The correct order to carry out operations. It stands for Brackets, Indices, Division, Multiplication, Addition, Subtraction.

 C

Capacity

How much something will hold. For example, a beaker with a capacity of 200 ml can hold 200 ml of liquid.

Certain

When something will definitely happen.

Circumference

The perimeter (distance around the outside) of a circle.

Clockwise

Movement in the same direction as the hands of a clock.

 D

Data

Another word for information.

Decimal Number

A number with a decimal point (.) in it. For example, 0.75.

Diameter

The distance from one side of a circle to the other, going straight through the middle. The diameter is twice the radius.

Dimension

A number that tells you about the size of an object. For example, its length.

Estimate

A close guess at what an answer will be.

Even Chance

When something is as likely to happen as it is not to happen.

Formula

A rule for working out an amount.

Fraction

A way of showing parts of a whole. For example, ¼ (one quarter).

Frequency Table

A tally chart with an extra column that shows the total of each tally (the frequencies).

Front Elevation

A 2D diagram to show how a 3D object looks from the front.

Function Machine

A way of showing formulas that have more than one step.

Impossible

When there's no chance at all of something happening.

Interest

Money paid or added on to a value over time, given as a percentage.

Length

How long something is. Length can be measured in different units, for example, millimetres (mm), centimetres (cm), or metres (m).

Likely

When something isn't certain, but there's a high chance it will happen.

Line Graph

A graph which shows data using a line.

Line of Symmetry

A shape with a line of symmetry has two halves that are mirror images of each other. If the shape is folded along this line, the two sides will fold exactly together.

Map Scale

A rule that tells you how far a given distance on a map is in real life.

Mean

A type of average. To calculate the mean you add up all the numbers and divide the total by how many numbers there are.

Mileage Chart

A type of table that shows you the distance between different places.

Mixed Number

When you have a whole number and a fraction together. For example, 2¼ (two and a quarter).

Negative Number

A number less than zero. For example, −2.

Net

A 3D shape folded out flat. You can use a net to help you make a 3D object. For example, you can use a net to make a box.

Percentage

A way of showing how many parts you have out of 100. For example, twenty percent (20%) is the same as 20 parts out of 100.

Perimeter

The distance around the outside of a shape.

Pie Chart

A circular chart that is divided into sections (that look like slices of a pie). The size of each section depends on how much or how many of something it represents.

Plan View

A 2D diagram to show how a 3D object looks from above.

Probability

The likelihood (or chance) of an event happening or not.

Protractor

A piece of equipment used to measure angles in degrees (°) up to 180°.

Quadrilateral

A 2D shape with 4 straight sides and 4 corners.

Radius

The distance from the side of a circle to the middle. The radius is half the diameter.

Range

The difference between the biggest and smallest numbers in a data set.

Ratio

A way of showing how many things of one type there are compared to another. For example, if there are 3 red towels to every 1 white towel then the ratio of red to white towels is 3:1.

Right Angle

A square corner.

Side Elevation

A 2D diagram to show how a 3D object looks from the side.

Square Number

A number multiplied by itself. For example, 5 squared (5^2) is the same as 5×5.

Symmetry

See line of symmetry.

Table

A way of showing data. In a table, data is arranged into columns and rows.

Tally Chart

A chart used for putting data into different categories. You use tally marks (lines) to record each piece of data in the chart.

Triangle

A 2D shape with 3 straight sides and 3 corners.

Unit

A way of showing what type of number you've got. For example, metres (m) or grams (g).

Unlikely

When something isn't impossible, but it probably won't happen.

Volume

The amount of space something takes up.

W

Weight

How heavy something is. Grams (g) and kilograms (kg) are common units for weight.

Index

M1ESRA1